Supporting Students' Development of Measuring Conceptions: Analyzing Students' Learning in Social Context

Supporting Students' Development of Measuring Conceptions: Analyzing Students' Learning in Social Context

Michelle Stephan
Purdue University Calumet

Janet Bowers
San Diego State University

Paul Cobb
Vanderbilt University

with

Koeno Gravemeijer
Freudenthal Institute

Neil Pateman, *Series Editor*
University of Hawaii

NCTM
NATIONAL COUNCIL OF
TEACHERS OF MATHEMATICS

Copyright © 2003 by
THE NATIONAL COUNCIL OF TEACHERS OF MATHEMATICS, INC.
1906 Association Drive, Reston, VA 20191-1502
(703) 620-9840; (800) 235-7566; www.nctm.org

All rights reserved

Library of Congress Cataloging-in-Publication Data

Supporting students' development of measuring conceptions : analyzing students' learning in social context / by Michelle Stephan, Janet Bowers, and Paul Cobb.
    p. cm. — (Journal for research in mathematics education. Monograph, ISSN 0883-9530 ; no. 12)
  ISBN 0-87353-556-1
1. Length measurement—Study and teaching (Primary)—Evaluation. 2. Mathematics—Study and teaching—Research—Evaluation. I. Stephan, Michelle. II. Bowers, Janet. III. Cobb, Paul. IV. Series.
  QA465.S78 2003
  372.7--dc22
                    2003023612

The publications of the National Council of Teachers of Mathematics present a variety of viewpoints. The views expressed or implied in this publication, unless otherwise noted, should not be interpreted as official positions of the Council.

Printed in the United States of America

Table of Contents

Acknowledgments vi

Chapter 1   Investigating Students' Reasoning About
Linear Measurement as a Paradigm Case of
Design Research 1
Paul Cobb, Vanderbilt University

Chapter 2   Reconceptualizing Linear Measurement Studies:
The Development of Three Monograph Themes 17
Michelle Stephan, Purdue University Calumet

Chapter 3   The Methodological Approach to
Classroom-Based Research 36
Michelle Stephan, Purdue University Calumet
Paul Cobb, Vanderbilt University

Chapter 4   A Hypothetical Learning Trajectory on Measurement
and Flexible Arithmetic 51
Koeno Gravemeijer, Freudenthal Institute
Janet Bowers, San Diego State University
Michelle Stephan, Purdue University Calumet

Chapter 5   Coordinating Social and Individual Analyses:
Learning as Participation in Mathematical Practices 67
Michelle Stephan, Purdue University Calumet
Paul Cobb, Vanderbilt University
Koeno Gravemeijer, Freudenthal Institute

Chapter 6   Continuing the Design Research Cycle:
A Revised Measurement and Arithmetic Sequence 103
Koeno Gravemeijer, Freudenthal Institute
Janet Bowers, San Diego State University
Michelle Stephan, Purdue University Calumet

Chapter 7   Conclusion 123
Michelle Stephan, Purdue University Calumet

Acknowledgments

The investigation reported in this monograph was supported by the National Science Foundation under grant number REC 9814898 and by the Office of Educational Research and Improvement (OERI) under grant number R305A60007. The opinions expressed do not necessarily reflect the views of either the Foundation or OERI.

The project team that engaged in the research reported in this monograph included Paul Cobb, Beth Estes, Kay McClain, Maggie McGatha, Beth Petty, and Michelle Stephan. Erna Yackel and Koeno Gravemeijer also collaborated on this project but did not participate on a daily basis.

Chapter 1

Investigating Students' Reasoning About Linear Measurement as a Paradigm Case of Design Research

Paul Cobb

The issues discussed in this monograph are situated in a description of a design experiment that focused on linear measurement and was conducted in a first-grade class. Because the focus on this particular mathematical domain might appear to be relatively narrow, I want to clarify at the outset that the report of this experiment is framed as a paradigm case in which to illustrate general features of the design-experiment methodology. Design experiments involve both developing instructional designs to support particular forms of learning and systematically studying those forms of learning within the context defined by the means of supporting them. This monograph also emphasizes that design experiments are conducted with a theoretical as well as a pragmatic intent, in that the goal is not merely to refine a particular instructional design. In the example at hand, the experiment was conducted with the intent of contributing to the development of a domain-specific instructional theory of linear measurement. A theory of this type specifies both a substantiated learning trajectory that aims at significant mathematical ideas and the demonstrated means of supporting learning along that trajectory. This concern for domain-specific theories reflects the view that the explanations and understandings inherent in them are essential if educational improvement is to be a long-term, generative process.

This attention to methodological issues is timely given the increasing prominence of the Design Research methodology in mathematics education. For example, Suter and Frechtling (2000) recently identified Design Research as one of the five primary research methodologies in mathematics and science education. As they note, about 20% of the research projects funded by the National Science Foundation's (NSF's) Division of Research, Evaluation, and Communication between 1996 and 1998 were design experiments. Not surprisingly, this increasing activity finds expression in the growing number of articles written by mathematics educators who report design experiments (e.g., Bowers & Nickerson, 2001; Lehrer, Jacobson, Kemney & Strom, 1999; Moss & Case, 1999; Rasmussen, 1998; Simon, 1995). However, as Suter and Frechtling go on to observe, a number of relatively common misconceptions about the methodology are apparent. Although the methodology is dynamic and open for adaptation, several central tenets serve as the basis for Design Research, namely, that instruction should (a) be experientially real for students, (b) guide students to reinvent mathematics using their commonsense experience, and (c) provide opportunities for students to create self-developed models. One of the goals of this monograph is to illustrate these tenets by grounding them in the concrete situation of a specific experiment. In this regard, the monograph complements three chapters on Design Research in a recent handbook on research design in mathematics and science education (cf. Cobb, 2000; Confrey & Lachance, 2000; Simon, 2000).

All the chapters in this monograph explore three themes that concern design researchers: (1) the relation between the development of instructional designs and the analysis of students' learning, (2) the relation between communal classroom processes and individual students' reasoning, and (3) the role of tools in supporting development. The first of these themes stems directly from the dual emphasis in Design Research on the development of domain-specific instructional theories and the formulation, testing, and revision of instructional designs. The relevance of the second theme is indicated by theoretical debates that have appeared in leading educational journals concerning the relation between collective communicational processes and the reasoning of the participating students (e.g. Anderson, Greeno, Reder, & Simon, 2000; Anderson, Reder, & Simon, 1996; Anderson, Reder, & Simon, 1997; Cobb & Bowers, 1999; Greeno, 1997; Lerman, 1996, 2000; Steffe & Thompson, 2000a). These exchanges indicate that no clear consensus has been reached on how the relation between these two types of processes might be productively conceptualized. We pursue that question and elaborate our contribution to the debate by exploring the interplay between individual students' reasoning and collective processes of communication and interaction as they played out in the case study in this monograph.

The importance of the third theme of mathematical tools becomes apparent once one notes that the development of tools to support students' learning is central to instructional design. The realization that students' mathematical learning is profoundly influenced by the tools they use, however, underpins a range of different approaches to instructional design. For example, Doerr (1995) distinguishes between expressive and exploratory approaches to design. In expressive approaches, students are encouraged to invent and test the adequacy of tools, such as notation systems, that express their developing understandings. In effective designs of this type, students develop increasingly sophisticated mathematical understandings by incrementally reformulating their informal knowledge (e.g., Bednarz, Dufour-Janvier, Portier, & Bacon, 1993; diSessa, Hammer, Sherin, & Kolpakowski, 1991). In contrast, exploratory approaches involve the development of computer environments in which students can investigate the links between conventional mathematical symbol systems and selected everyday phenomena. For example, Kaput (1994) and Nemirovsky (1994) have both developed environments in which distance-time graphs are introduced at the outset and students explore the use of these graphs as a means of representing various types of motions. In effective designs of this type, the computer environments span the initial gulf between students' authentic experiences and mathematical ways of symbolizing, such as graphing. The comparison of the expressive and exploratory approaches highlights a tension inherent in instructional design between the ideal of building on students' contributions and the need to decide in advance the tools and symbolizations that students should eventually come to use. This monograph addresses this tension when describing the instructional design developed and revised in the course of the measurement experiment.

In the following pages, I first situate the three themes of the monograph by giving a brief historical overview of Design Research. In the remainder of this introductory chapter, I focus on two main aspects of the Design Research methodology. These aspects concern the instructional theory that undergirds the formulation of instructional designs and the interpretive framework that guides the analysis of classroom events.

EXPERIMENTING IN THE CLASSROOM

As I have indicated, the intimate relation between theory and practice is one of the defining characteristics of the Design Research methodology. On the one hand, instructional design serves as a primary context for research and the development of theory. On the other hand, analyses of students' learning and the means by which it was supported inform researchers in their revision of the instructional design. As a consequence, the purpose when experimenting in the classroom is not to try to demonstrate that the initial design formulated in advance of the experiment works. Instead, the purpose is to test and improve the initial design as guided by both ongoing and retrospective analyses of classroom activities and events.

Methodologies in which instructional design serves as a context for the development of theories of learning and instruction have a long history, particularly in the former Soviet Union (Menchinskaya, 1969). The term *Design Research,* however, was coined relatively recently and is most closely associated with Ann Brown (1992). In her formulation, Design Research emerged as a reaction against traditional approaches to the study of learning that emphasize the control of variables. In a more recent discussion of the methodology, Collins (1999) also takes traditional psychological methodology as his point of reference. For example, he follows Brown in contrasting studies of learning conducted in relatively artificial laboratory settings with design experiments that focus on learning as it occurs in complex and messy settings, such as classrooms. The purpose of Design Research for Brown and Collins, however, is not merely to study learning in situ. They emphasize that the methodology is highly interventionist and involves developing designs that embody testable conjectures about the means of supporting an envisioned learning process. In addition, they underscore that the methodology has both a theoretical and a pragmatic intent by drawing an analogy with design sciences, such as aeronautical engineering. As they note, an aeronautical engineer creates a model, subjects it to certain stresses, and generates data to test and revise theoretical conjectures inherent in the model. Similarly, a Design Research team creates a design to support an envisioned learning process, conducts an experiment to subject the design to certain stresses, and generates data to test and revise theoretical conjectures derived from prior research that are inherent in the design. This analogy also clarifies that in both aeronautical engineering and Design Research, generalization is accomplished by means of an explanatory framework rather than by means of a representative sample, in that the theoretical insights and understandings developed during one or more experiments can feed forward to influence the analysis of events and thus pedagogical planning and decision making in other classrooms (cf. Steffe & Thompson, 2000b).

The concerns for the development of domain-specific instructional theories and for potential generalizability serve to differentiate Design Research from the activity of thoughtful practitioners who continually seek to improve their practices. The parallels are indicated by Franke, Carpenter, Levi, & Fennema's (2001) observation that mathematics teaching is for some teachers a knowledge-generating activity in the course of which they elaborate and refine their understandings of both their students' reasoning and the means of supporting its development. Whereas the teachers described by Franke and others work in their local classroom and school settings to improve their effectiveness in supporting their students' learning, Design Researchers explicitly frame aspects of the learning processes they are attempting to support as paradigm cases of broader classes of phenomena. Thus, whereas practitioners are primarily concerned with the effectiveness of their practices in their local settings, Design Researchers also focus on the development of theoretical insights when they plan for an experiment, generate data during an experiment, and conduct retrospective analyses.

Collins's (1999) and Brown's (1992) works have proved to be foundational to the emerging field of the learning sciences, itself a direct descendent of cognitive science (cf. DeCorte, Greer, & Verschaffel, 1996). As the increasing adoption of their terminology indicates, their arguments have also had a considerable influence in mathematics education. Although Design Research in the learning sciences and that in mathematics education are highly compatible, their histories differ. The emergence of the learning sciences from cognitive science signaled a relatively radical change of priorities, whereas the development of Design Research in mathematics education has been more evolutionary and builds on two existing research traditions: the constructivist teaching experiment and Realistic Mathematics Education developed at the Freudenthal Institute in the Netherlands.

The constructivist teaching experiment methodology was developed by Steffe and his colleagues (Cobb & Steffe, 1983; Steffe, 1983; Steffe & Kieren, 1994; Steffe & Thompson, 2000b). The purpose of the teaching experiment as formulated by Steffe and his colleagues is to enable researchers to investigate the *process* by which individual students reorganize their mathematical ways of knowing. To this end, a researcher typically interacts with a student one-on-one and attempts to precipitate his or her learning by posing judiciously chosen tasks and by asking follow-up questions, often with the intention of encouraging the student to reflect on his or her mathematical activity. The primary products of a teaching experiment of this type typically consist of conceptual models composed of theoretical constructs developed to account for the learning of the participating student(s) (Thompson & Saldanha, 2000). The intent in developing these models is that they will prove useful when accounting for the learning of other students and can thus inform teachers in their decision making. Although the researcher acts as a teacher in this methodological approach, the primary emphasis of the constructivist teaching experiment is on the interpretation of students' mathematical reasoning rather than on the development of instructional designs.

An initial attempt to adapt the constructivist teaching experiment methodology to the classroom setting involved creating a complete set of instructional activities for second- and third-grade classrooms (Cobb, Yackel & Wood, 1989). Nonetheless, the primary focus of these experiments was consistent with Steffe's (1983) emphasis on the development of explanatory constructs rather than the formulation of instructional designs. In particular, Cobb and his colleagues gave priority to developing an interpretive framework that would enable them to situate students' mathematical learning within the social context of the classroom (cf. Cobb & Yackel, 1996). In the course of this work, they came to view their lack of positive design heuristics that could guide the development of instructional activities as a severe limitation. In general, constructivism and related theories present a number of negative heuristics that rule out a range of approaches to instructional design. The types of cognitive and interactional analyses that Cobb and others conducted at that time, however, did not provide an adequate basis for design, because the analyses did not focus explicitly on the means by which students' mathematical learning was supported and organized.

The second research tradition on which Design Research is built, that of Realistic Mathematics Education (RME), offsets this weakness by focusing primarily on design rather than the development of explanatory theoretical constructs (cf. Gravemeijer, 1994; Streefland, 1991; Treffers, 1987). Guided by Freudenthal's (1973) notion of mathematics as a human activity and building on his didactical phenomenology of mathematics, RME researchers have developed, tried out, and modified instructional sequences in a wide range of mathematical domains. Further, as Treffers (1987) documents, reflection on this process of developing, testing, and revising specific instructional sequences has resulted in the delineation of a series of heuristics for instructional design in mathematics education. Gravemeijer (1994), for his part, has analyzed the process of the emergence of both these general design heuristics and the domain-specific instructional theories that constitute the rationale for particular instructional sequences. As will become apparent, the development of the measurement instructional sequence that is the focus of this monograph was guided by RME design heuristics. In addition, the processes by which the initial design of the instructional sequence was tested and revised during the first-grade design experiment is consistent with Gravemeijer's analysis of the development of a domain-specific instructional theory.

This brief historical account of Design Research should clarify that my colleagues and I attribute great importance both to the *instructional theory* that serves to orient the development of specific designs and to the *interpretive framework* that serves to organize classroom analyses that in turn influence the ongoing design effort. The remainder of this introduction focuses on these two core aspects of classroom Design Research.

DESIGNING LEARNING ENVIRONMENTS

Gravemeijer, Bowers, & Stephan (chapter 4 of this monograph) set the stage for their discussion of the measurement design experiment by giving an overview of RME design theory. Against this background, they then describe the conjectured learning trajectory that was formulated in advance of the experiment. As they make clear, the conjectured means of supporting the students' learning were not limited to the instructional tasks but included the students' use of several different types of tools for measuring. They emphasize that these tools were not designed to serve merely as a means by which students might express their reasoning. Instead, the design team conjectured that students might reorganize their reasoning as they used the tools. The approach taken to instructional design in the measurement-design experiment is therefore compatible with the proposition that the use of tools can serve not merely to amplify but to reorganize activity (cf. Dörfler, 1993; Pea, 1993). Stephan (chapter 2 of this monograph) emphasizes this point of agreement with distributed theories of intelligence by speaking of the students' *reasoning with* the tools both when describing the conjectured learning trajectory and when analyzing classroom events.

In addition to focusing on instructional tasks and associated resources, such as tools, the design team also considered potential means of support that typically fall beyond the purview of curriculum developers when preparing for the measurement experiment. This work included conjectures about both classroom discourse and the classroom activity structure as means of support. In the case of classroom discourse, these conjectures concerned norms or standards for what would count as an acceptable explanation of measuring activity. A distinction that Thompson, Philipp, Thompson, and Boyd (1994) make between calculational and conceptual orientations in teaching proved to be particularly relevant. The design team extended this distinction by developing a contrast between *calculational discourse* and *conceptual discourse*. The standards of argumentation in calculational discourse are such that an acceptable explanation need only describe the method or process by which a result is produced. With regard to measuring, a calculational explanation involves demonstrating how a measuring tool has been used to find the length of an object. An important point to clarify is that calculational discourse is not restricted to conversations that focus on the procedural manipulation of conventional tools and symbols whose use is a rule-following activity for students. The methods for producing results that students explain as they contribute to such a conversation, might, in fact, be self-generated and involve relatively sophisticated mathematical understandings. As a consequence, the contrast between calculational and conceptual discourse should not be confused with Skemp's (1976) well-known distinction between instrumental and relational understanding.

The reference to Skemp's (1976) work serves to emphasize that the defining characteristic of calculational discourse concerns the norms or standards for judging what counts as an acceptable argument rather than for judging the quality of students' understandings. In calculational discourse, contributions are acceptable if students describe how they produced a result, and they are not obliged to explain why they used a particular method. By this criterion, examples of classroom discourse presented in the literature to illustrate instruction compatible with current reform recommendations are, in fact, calculational rather than conceptual in nature. In contrast with this emphasis on methods or solution strategies, the issues that emerge as topics of conversation in conceptual discourse also include the reasons for calculating in particular ways. In measuring, these reasons concern the way in which particular methods of measuring structure the space measured into units of length.

I can best illustrate the distinction between these two types of classroom discourse by foreshadowing Stephan, Cobb, & Gravemeijer's analysis (chapter 5 of this monograph) of an incident that occurred early in the measurement design experiment. The teacher and observing researchers noted that different groups of students developed two different methods when they measured various distances by pacing heel to toe. In the ensuing whole-class discussion, the teacher asked selected children to measure the length of a rug in the classroom so that she could highlight the difference between these two methods. In one method, the student places one foot in line with the beginning of the rug but does not begin counting paces until she places the second foot heel to toe in front of the first (i.e., from our standpoint, the child fails to count her first pace). We might reasonably speculate that these students were counting their physical acts of placing down a foot per se instead of attempting to find the number of paces they needed to take to cover the linear extension of the rug. In the second method, the child counts her initial placement of a foot in line with the beginning of the rug as "one" and then continues by counting successive paces as "two, three, ..." The project team conjectured that these students might have been partitioning the linear extension of the rug instead of merely counting their physical movements per se.

If calculational norms had been established for argumentation, then justifications in which students repeatedly demonstrated to one another how they had counted their paces would have been acceptable. Imagine for a moment how students who failed to count the first pace might have interpreted calculational explanations of the other method. Such explanations would likely not have made sense to these students, because the method is unrelated to the task as they understood it: to count physical acts of pacing. These students might, however, have realized that the method being explained is more valued than their own and adjusted their approach accordingly. In doing so, they would have adopted the desired method but without reconceptualizing the nature of their activity. The crucial point to note is that a calculational discussion of the two methods provides little support for substantive mathematical learning. Instead, the students largely on their own would have to reconceptualize their activity as one of covering a distance.

A calculational exchange of this type can be contrasted with a conceptual discussion in which the reasons for counting paces in a particular way become an explicit topic of conversation. For example, a student who counted her placement of the first foot might explain that her paces completely fill the spatial extension of the rug without any overlaps and without leaving any gaps. Further, the students who used this method might argue that the students who did not count the first pace missed a piece of the rug. In the context of such an exchange, the way in which the two methods structure the spatial extension of the rug can emerge as an explicit criterion for comparing them. The emergence of this criterion, in turn, serves to support the reconceptualization of pacing as a distance-covering activity. Students' engagement in conceptual discussions of this type therefore provides them with resources that might enable them to reorganize their thinking. As Stephan and her colleagues demonstrate in their analysis of the measurement design experiment in chapter 5, these resources were not limited to what was said but include the notational schemes that were developed to record the ways in which different methods of measuring structured the linear extension of the object being measured. The goals for students' learning that were formulated when preparing for the design experiment were not restricted to methods of measuring but included the meaning of measuring (i.e., the partitioning of the linear extension of objects into units). As a consequence, students' engagement in conceptual discussions in which differing ways of structuring distance came to the fore was viewed as a primary means of advancing the pedagogical agenda.

To this point, I have discussed three means of supporting students' learning that were considered when planning for the measurement experiment: (1) the instructional activities, (2) the tools that students would use to measure, and (3) the nature of classroom discourse. The classroom activity structure constituted the final means of support to which the design team attended prior to the experiment. The proposed organization of classroom activities had three main phases:

1. The teacher would develop with the students an ongoing narrative in which the characters in the narrative encounter problems that involve either measuring with an existing tool or developing a new measuring tool.
2. The students would engage in measuring activities either individually or in pairs.
3. The teacher would capitalize on the range of interpretations and solutions that the students developed during individual or small-group work to lead a whole-class discussion in which mathematically significant ideas that advance the pedagogical agenda emerge as topics of conversation.

The design team conjectured that students' engagement in the first of these three phases would support their learning in two specific ways. Both these conjectures were premised on the assumption that students would identify with the characters in the narrative as they attempted to resolve the various problems they encountered. The first conjecture was that students' measuring activity would take on a broader significance and purpose within the context of the narrative so that they would not merely be measuring at the teacher's behest. The design team's second conjecture was that if students identified with the characters in the narrative, the need to develop various new measuring tools as formulated within the narrative would seem plausible to them. Further, a discussion of possible solutions would serve to clarify the design specifications for the new tool. As a consequence, new tools would not seem arbitrary to students even on those occasions when the teacher introduced them. Instead, students would view tools introduced in this way as reasonable solutions to problems that had significance within the context of the narrative.

These two conjectured ways in which the ongoing narrative might support students' learning constituted the primary justification for the first phase of the proposed classroom-activity structure. The issues I raised when clarifying the distinction between calculational and conceptual discourse serve to justify the last two phases of the classroom-activity structure. Clearly the intent of the whole-class discussions was not simply to provide students with an occasion to share their reasoning. Instead, the design team's overriding concern was with the quality of the discussions as social events in which students would participate. From this perspective, the value of whole-class discussions is suspect unless mathematically significant issues that advance the instructional agenda emerge as explicit topics of conversation. Conversely, students' participation in substantive, conceptual discussions is viewed as a primary means of supporting their enculturation into the values, beliefs, and ways of knowing of the discipline.

This discussion of the classroom activity structure serves to clarify the last of the four means of support considered when preparing for the measurement experiment:

1. The instructional tasks
2. The tools the students would use to measure
3. The nature of classroom discourse
4. The classroom activity structure

Because I have focused on each means of support separately, they might be construed as a largely independent set of factors that influence learning separately. An important point to emphasize is that the design team viewed the four means of support as highly interrelated. For example, the instructional tasks as actually realized in the classroom depended on the extent to which the students identified with the characters in the narratives, the tools they used to measure, and the nature of the classroom discussions. One can easily imagine, for example, how the tasks might be realized differently if a conventional measuring tool, such as a ruler, were introduced at the outset or if no whole-class discussions were held and the teacher simply graded the accuracy of the students' measuring activity.

In light of these interdependencies, one can reasonably view the various means of support as constituting a single classroom activity system. This perspective is compatible with Stigler and Hiebert's (1999) contention that teaching should be viewed as a system. In making this claim, Stigler and Hiebert directly challenge analyses that decompose teachers' instructional practices into a number of independent moves or competencies. They instead propose that the meaning and significance of any particular facet of a teacher's instructional practice becomes apparent only when analyzed within the context of the entire practice. In a similar manner, the four means of support considered when planning for the measurement design experiment should be viewed as aspects of a single classroom activity system. Instructional design from this point of view therefore involves designing classroom activity systems in which students develop significant mathematical ideas as they participate in their exploration and contribute to their evolution.

The orientation to design that we took when preparing for the measurement experiment sits uncomfortably with approaches that focus exclusively on tasks and tools. In these formulations, tasks are typically cast as the cause and learning as the effect. An important point to emphasize is that when Gravemeijer and his colleagues discuss tasks and measuring tools while outlining the proposed instructional sequence in chapter 4, they are describing aspects of an envisioned classroom activity system. Their overview of the proposed instructional sequence is made against the backdrop of assumptions about both the classroom activity structure and the classroom discourse. As a consequence, their conjectures about students' learning as they use particular tools to complete particular tasks are metonymies for conjectures about students' learning as they participate in an envisioned classroom activity system.

ANALYZING MATHEMATICAL LEARNING IN SOCIAL CONTEXT

As I noted in the first part of this introduction, the design of classroom learning environments is one of the two principal aspects of Design Research. The second aspect concerns the analysis of mathematical learning as situated within the social context of the classroom. In chapter 3 Stephan and Cobb describe in some detail the interpretive framework that was used to organize the analysis of the measurement teaching experiment. Rather than summarize their presentation by discussing the constructs that compose this framework, I instead attempt to clarify its rationale. As should become apparent, the framework can best be viewed as a potentially revisable solution to problems and issues encountered while conducting design experiments.

As a first step in outlining the rationale, a point worth reiterating is that classrooms are complex, messy, and sometimes confusing places. One of the concerns that all Design Researchers need to address is that of developing an analytic framework that enables research participants to discern pattern and order in what often appear to be ill-structured events. These concerns and interests give rise to several criteria that an analytical approach should satisfy if it is to contribute to reform in mathematics education as an ongoing, iterative process of continual improvement. These criteria include the following:

- The results from the analyses should feed back to improve the instructional designs.
- The methodology should permit documentation of the collective mathematical learning of the classroom community over the extended periods of time spanned by design experiments.
- The analysis should permit documentation of the developing mathematical reasoning of individual students as they participate in communal classroom processes.

The first of these criteria follows directly from the emphasis on testing and revising the conjectures inherent in initial designs when conducting a design experiment. The second criterion, which emphasizes the importance of focusing on the mathematical learning of the classroom community, stems from the approach to instructional design that I have outlined. As noted, the designer develops conjectures about an anticipated learning trajectory when preparing for a design experiment. These conjectures, however, cannot be about the trajectory of each and every student's learning for the straightforward reason that significant qualitative differences are present in their mathematical thinking at any point in time. As a consequence, descriptions of planned instructional approaches written to imply that all students will reorganize their thinking in particular ways at particular points in an instructional sequence involve, at best, questionable idealizations. For similar reasons, analyses that speak of changes in students' reasoning are potentially misleading in that they imply that students will all reorganize their thinking in the same way. I should acknowledge that I have, in fact, spoken in these terms to this point in this introduction for ease of explication. Although this approach proved to be adequate when discussing the various means of supporting and organizing learning, it does not offer the precision we need to improve our designs.

Against the background of these considerations, an issue that my colleagues and I have sought to address is that of clarifying what the envisioned learning trajectories central to our, and others', work as instructional designers might be about. The resolution that we propose involves viewing a hypothetical learning trajectory as consisting of conjectures about the collective mathematical development of the classroom community. This proposal, in turn, indicates the need for a theoretical construct that enables us to talk explicitly about collective mathematical learning, and for this reason we have developed the notion of a classroom mathematical practice. Stephan and her colleagues (chapters 3 and 5 of this monograph) define and illustrate this construct when they present their analysis of the measurement design experiment. For the present, it suffices to note that a hypothetical learning trajectory that is framed in these terms consists of an envisioned sequence of mathematical practices together with conjectures about the means of supporting and organizing the emergence of each practice from prior practices.

The last of the three criteria that an interpretive approach should satisfy focuses on qualitative differences in individual students' mathematical reasoning. The rationale for this criterion is again deeply rooted in the activity of experimenting in classrooms. The classroom activity structure that I described when discussing our preparations for the measurement experiment is representative in that, in most of the design experiments that I and my colleagues have conducted, the students work either individually or in small groups before convening for a whole-class discussion of their interpretations and solutions. During the individual or small-group work, the teacher and one or more of the project staff normally circulate in the classroom to gain a sense of the diverse ways in which students are interpreting and solving the instructional activities. Toward the end of this phase of the lesson, the teacher and project staff members confer briefly to prepare for the whole-class discussion. In doing so, they routinely focus on the qualitative differences in the students' reasoning to develop conjectures about mathematically significant issues that might, with the teacher's guidance, emerge as topics of conversation. They might, for example, conjecture that a particular mathematical issue will emerge if two specific types of solutions are compared during the discussion. Given this pragmatic focus on individual students' reasoning, we require an analytic approach that takes account of the diverse ways in which students participate in communal classroom practices. In the hands of a skillful teacher, this diversity can, in fact, be a primary resource on which the teacher can capitalize to support the collective mathematical learning of the classroom community.

This discussion of the three criteria that an interpretive framework should satisfy illustrates the pragmatic orientation we take in viewing theoretical constructs as conceptual tools whose development reflects particular interests and concerns. From this perspective, the relevant concern when assessing the value of theoretical constructs is whether they enable us to be more effective in supporting students' mathematical learning. Clearly, also, an interpretive approach that satisfies the three criteria will characterize students' mathematical learning in situated terms. In doing so, however, an interpretive approach will take a different view of the classroom activity system. The aspects of this system on which I focused when discussing the various means of supporting and organizing mathematical learning dealt with resources (e.g., tasks, tools) and with characteristics of collective activity (e.g., the structure of classroom activities, classroom discourse). Missing from this picture is an indication of the specific processes by which increasingly sophisticated mathematical ways of reasoning might emerge as students participate in these collective activities by using the tools to complete instructional tasks. The interpretive framework that Stephan and Cobb (chapter 3 of this monograph) employ when analyzing the data generated in the course of the measurement design experiment is designed to address this limitation.

CONCLUSION

My purpose in this introduction has been to situate the measurement design experiment within a broader methodological context. To this end, I highlighted cycles of design and analysis as a defining characteristic of Design Research. I also noted the pragmatic emphasis of the methodology in addressing problems of supporting students' mathematical learning similar to those addressed by practitioners. In addition, I discussed the theoretical intent of developing domain-specific instructional theories that enable the results of an experiment to be generalized by means of an explanatory framework. This dual focus on theoretical issues and pragmatic concerns makes the methodology particularly well suited to the task of investigating the prospects and possibilities for reform in mathematics education at the classroom level.

The two core aspects of Design Research on which I have focused are the design theory that guides the development of specific designs and the interpretive framework that guides the analysis of classroom events. In discussing the design theory, I emphasized that the conjectures developed when preparing for the measurement experiment were not restricted to curriculum developers' traditional focus on tasks and tools. As I illustrated, the conjectures also took account of both the nature of classroom discourse and the classroom activity structure. To accommodate this broader perspective on design, I introduced the notion of the classroom activity system. The hallmark of such systems is that they are intentionally designed to produce the learning of significant mathematical ideas as students participate in them and contribute to their evolution. In light of this shift from a focus on tasks and tools to a broader concern for the activity systems that constitute the social situations of students' learning, one might appropriately speak of design experiments rather than of teaching experiments. The term *teaching experiment* indicates that the methodology is agenda driven and highly interventionist. The term *design experiment* adds a theoretical dimension by indicating the broader perspective from which the interventions are conceptualized.

My discussion of the second core aspect of Design Research focused on the three criteria that an appropriate interpretive framework should satisfy. As I noted, the resulting accounts of students' mathematical learning are necessarily situated in that they are tied to analyses of the actual environment in which that learning occurred. As Stephan and others' analysis in chapter 5 of the measurement experiment illustrates, one can therefore disentangle aspects of this environment that served to support the development of the students' reasoning. Doing so, in turn, leads to the development of testable conjectures about the way in which those means of support and thus the instructional design might be improved. In this regard, the methodology employed by Stephan and her colleagues remains true to Brown's (1992) and Collins's (1999) vision of educational research as a process of ongoing, iterative improvement.

Given my purpose in this chapter, I have focused on general characteristics of the design experiment rather than on concrete aspects of the methodology. These aspects are discussed in the remainder of this monograph and include the specification of the relation between classroom-based research and instructional design, the interpretive framework used to analyze learning in the social context of the classroom, and the role of tools in both learning and design. In chapter 2, Stephan synthesizes the literature on students' measurement conceptions to further develop the three monograph themes. In chapter 3, Stephan and Cobb discuss the interpretive framework and method that guided the analysis of students' learning in the social context of the classroom. In chapter 4, Gravemeijer, Bowers, and Stephan illustrate the central instructional design heuristics of RME and present a hypothetical learning trajectory that served as the basis for the measurement experimentation. In chapter 5, Stephan, Cobb, and Gravemeijer offer an analysis that coordinates the learning of the classroom community with the reasoning of the participating students. Gravemeijer, Bowers, and Stephan then use this analysis in chapter 6 to influence the revision of the prior hypothetical learning trajectory. Finally, in chapter 7, Stephan revisits the three monograph themes in light of the measurement experiment. The content of the chapters was organized in this manner to illustrate the cyclic process of instructional design and classroom-based research that is characteristic of Design Research.

REFERENCES

Anderson, J., Greeno, J., Reder, L., & Simon, H. (2000). Perspectives on learning, thinking, and activity. *Educational Researcher, 29*(4), 11–13.

Anderson, J., Reder, L., & Simon, H. (1996). Situated learning and education. *Educational Researcher, 25*(4), 5–11.

Anderson, J., Reder, L., & Simon, H. (1997). Situative versus cognitive perspectives: Form versus substance. *Educational Researcher, 26*(1), 18–21.

Bednarz, N., Dufour-Janvier, B., Portier, L., & Bacon, L. (1993). Socioconstructivist viewpoint on the use of symbolism in mathematics education. *The Alberta Journal of Educational Research, 39*(1), 41–58.

Bowers, J., & Nickerson, S. (2001). Documenting the development of a collective conceptual orientation in a college-level mathematics course. *Mathematical Thinking and Learning, 3,* 1–28.

Brown, A. L. (1992). Design experiments: Theoretical and methodological challenges in creating complex interventions in classroom settings. *Journal of the Learning Sciences, 2,* 141–178.

Cobb, P. (2000). Conducting teaching experiments in collaboration with teachers. In A. E. Kelly & R. A. Lesh (Eds.), *Handbook of research design in mathematics and science education* (pp. 307–334). Mahwah, NJ: Erlbaum.

Cobb, P., & Bowers, J. (1999). Cognitive and situated perspectives in theory and practice. *Educational Researcher, 28*(2), 4–15.

Cobb, P., & Steffe, L. P. (1983). The constructivist researcher as teacher and model builder. *Journal for Research in Mathematics Education, 14,* 83–94.

Cobb, P. & Yackel, E. (1996). Constructivist, emergent, and sociocultural perspectives in the context of developmental research. *Educational Psychologist, 31,* 175–190.

Cobb, P., Yackel, E., & Wood, T. (1989). Young children's emotional acts while doing mathematical problem solving. In D. B. McLeod & V. M. Adams (Eds.), *Affect and mathematical problem solving: A new perspective* (pp. 117–148). New York: Springer-Verlag.

Collins, A. (1999). The changing infrastructure of educational research. In E. C. Langemann & L. S. Shulman (Eds.), *Issues in education research* (pp. 289–298). San Francisco: Jossey Bass.

Confrey, J., & Lachance, A. (2000). Transformative reading experiments through conjecture-driven research design. In A. E. Kelly & A. Lesh (Eds.), *Handbook of research design in mathematics and science education* (pp. 231–266). Mahwah, NJ: Erlbaum.

DeCorte, E., Greer, B., & Verschaffel, L. (1996). Mathematics learning and teaching. In D. Berliner & R. Calfee (Eds.), *Handbook of educational psychology* (pp. 491–549). New York: Macmillan.

diSessa, A. A., Hammer, D., Sherin, B., and Kolpakowski, T. (1991). Inventing graphing: Meta-representational expertise in children. *Journal of Mathematical Behavior, 10,* 117–160.

Doerr, H. M. (1995, April). *An integrated approach to mathematical modeling: A classroom study.* Paper presented at the Annual Meeting of the American Educational Research Association, San Francisco, CA.

Dörfler, W. (1993). Computer use and views of the mind. In C. Keitel & K. Ruthven (Eds.), *Learning from computers: Mathematics education and technology*. Berlin, Germany: Springer-Verlag.

Franke, M. L., Carpenter, T. P., Levi, L., & Fennema, E. (2001). Capturing teachers' generative change: A follow-up study of teachers' professional development in mathematics. *American Educational Research Journal, 38*, 653–689.

Freudenthal, H. (1973). *Mathematics as an educational task*. Dordrecht, Netherlands: Reidel.

Gravemeijer, K. (1994). *Developing realistic mathematics education*. Utrecht, Netherlands: CD-ß Press.

Greeno, J. G. (1997). On claims that answer the wrong questions. *Educational Researcher, 26*(1), 5–17.

Kaput. J. J. (1994). The representational roles of technology in connecting mathematics with authentic experience. In R. Biehler, R. W. Scholz, R. Sträßer, & B. Winkelmann (Eds.), *Didactics of mathematics as a scientific discipline* (379–397). Dordrecht, Netherlands: Kluwer.

Lehrer, R., Jacobson, C., Kemney, V., & Strom, D. A. (1999). Building upon children's intuitions to develop mathematical understanding of space. In E. Fennema & T. R. Romberg (Eds.), *Mathematics classrooms that promote understanding* (pp. 63–87). Mahwah, NJ: Erlbaum.

Lerman, S. (1996). Intersubjectivity in mathematics learning: A challenge to the radical constructivist paradigm? *Journal for Research in Mathematics Education, 27*, 133–150.

Lerman, S. (2000). A case of interpretations of "social": A response to Steffe and Thompson. *Journal for Research in Mathematics Education, 31*(2), 210–227.

Menchinskaya, N. A. (1969). Fifty years of Soviet instructional psychology. In J. Kilpatrick & I. Wirzup (Eds.), *Soviet studies in the psychology of learning and teaching mathematics* (Vol. 1). Stanford, CA: School Mathematics Study Group.

Moss, J., & Case, R. (1999). Developing children's understanding of the rational numbers: A new model and an experimental curriculum. *Journal for Research in Mathematics Education, 30*, 122–147.

Nemirovsky, R. (1994). On Ways of Symbolizing: The Case of Laura and Velocity Sign. *Journal of Mathematical Behavior, 13*, 389–422.

Pea, R. D. (1993). Practices of distributed intelligence and designs for education. In G. Salomon (Ed.), *Distributed cognitions* (pp. 47–87). New York: Cambridge University Press.

Rasmussen, C. (1998). *Reform in differential equations: A case study of students' understanding and difficulties*. Paper presented at the annual meeting of the American Educational Research Association, San Diego, CA.

Simon, M. A. (1995). Reconstructing mathematics pedagogy from a constructivist perspective. *Journal for Research in Mathematics Education, 26*, 114–145.

Simon, M. A. (2000). Research on the development of mathematics teachers: The teacher development experiment. In A. E. Kelly & R. A. Lesh (Eds.), *Handbook of research design in mathematics and science education* (pp. 335–359). Mahwah, NJ: Erlbaum.

Skemp, R. (1976). Relational understanding and instrumental understanding. *Mathematics Teacher, 77*, 20–26.

Steffe, L. P. (1983). The teaching experiment methodology in a constructivist research program. In M. Zweng, T. Green, J. Kilpatrick, H. Pollak, & M. Suydam (Eds.), *Proceedings of the Fourth International Congress on Mathematical Education*. Boston: Birkhauser.

Steffe, L. P., & Kieren, T. (1994). Radical constructivism and mathematics education. *Journal for Research in Mathematics Education, 25*, 711–733.

Steffe, L. P., & Thompson, P. (2000a). Interaction or intersubjectivity? A reply to Lerman. *Journal for Research in Mathematics Education, 31*(2), 191–209.

Steffe, L. P., & Thompson, P. W. (2000b). Teaching experiment methodology: Underlying principles and essential elements. In A. E. Kelly & R. A. Lesh (Ed.), *Handbook of research design in mathematics and science education* (pp. 267–307). Mahwah, NJ: Erlbaum.

Stigler, J. W., & Hiebert, J. (1999). *The teaching gap*. New York: Free Press.

Streefland, L. (1991). *Fractions in realistic mathematics education. A paradigm of developmental research*. Dordrecht, Netherlands: Kluwer.

Suter, L. E., & Frechtling, J. (2000). *Guiding principles for mathematics and science education research methods: Report of a workshop*. Washington, DC: National Science Foundation.

Thompson, A. G., Philipp, R. A., Thompson, P. W., & Boyd, B. (1994). Calculational and conceptual orientations in teaching mathematics. In Douglas B. Aichele (Ed.), *1994 Yearbook of the National Council of Teachers of Mathematics* (pp. 79–92). Reston, VA: National Council of Teachers of Mathematics.

Thompson, P. W., & Saldanha, L. A. (2000). Epistemological analyses of mathematical ideas: A research methodology. In M. Fernandez (Ed.), *Proceedings of the Twenty-Second Annual Meeting of the North American Chapter of the International Group for the Psychology of Mathematics Education* (pp. 403–407). Columbus, OH: ERIC Clearinghouse for Science, Mathematics, and Environmental Education.

Treffers, A. (1987). *Three dimensions: A model of goal and theory description in mathematics instruction—The Wiskobas Project.* Dordrecht, Netherlands: Reidel.

Chapter 2

Reconceptualizing Linear Measurement Studies: The Development of Three Monograph Themes

Michelle Stephan
Purdue University Calumet

Measuring lengths of objects is a common practice in a wide range of out-of-school situations. For example, some families keep track of their children's growth by measuring their heights and either keeping a written record or making physical marks on a wall. In addition, learning to measure forms a foundation for investigating other mathematical topics, such as proportion, decimals, and fractions, to name a few (cf. National Council of Teachers of Mathematics [NCTM], 2003). For these reasons, instruction in linear measurement is included in most elementary curricula. The NCTM *Principles and Standards for School Mathematics* (2000) emphasizes the importance of establishing a firm foundation in the underlying concepts and skills of measurement. The document emphasizes that children need to "understand the attributes to be measured as well as what it means to measure" (p. 103). A large body of literature on children's conceptions of measurement has amassed over the past three decades, with undoubtedly the most influential work being that of Jean Piaget and his colleagues (e.g., Piaget, Inhelder, Szeminska, 1960). Piaget and his colleagues identified developmental stages through which they claimed children pass as they learn to measure. As a result of this analysis, many researchers have tried to isolate the ages at which children develop certain measurement concepts. Other researchers have devised training programs to increase the acquisition rate of measurement concepts. Whereas each of these studies took a primarily individualistic approach to learning and training, few studies have been conducted that address social[1] influences on children's development of measuring abilities. As such, I argue that current measurement studies should include social aspects as well as tool use as an integral part of learning to measure.

---

[1] We use the term *social* throughout the monograph to refer to the interactions between the teacher and the students at the local level of the classroom, not to any particular societal characteristic, such as a student's racial background or SES status.

The purpose of this chapter is to identify and develop three research themes that drive the organization and substance of this monograph in regard to supporting students' development of linear measurement conceptions. To provide rationale for these three research themes, I first include an extensive literature review describing prior research on children's measurement conceptions. This review includes descriptions of studies conducted by Piaget and his colleagues and of training programs that were designed to support children's development of measurement skills. Second, I use this literature review as background to elaborate the three research themes that serve as the basis of inquiry for this study. Third, I elaborate our theoretical position on each of the research themes to argue that new investigations of children's measurement development should attend to social aspects of learning to measure, the use of tools to support development, and learning as it is closely tied to the designer's in-action decisions.

MEASUREMENT INVESTIGATIONS

Most investigations into children's conceptions of measurement have been based on studies that were published in the 1950s and 1960s by Piaget and his colleagues (i.e., Piaget & Inhelder, 1956; Piaget et al., 1960). Other researchers have set out to either substantiate or disprove the claims that Piaget made about children's development. Because the majority of the literature on children's conceptions of measurement focused on Piaget's stage theory, I begin this section by discussing Piaget's investigations on measurement. Further, I draw connections between our findings and those of Piaget throughout the monograph, in particular in chapter 5.

Piaget's Studies

Piaget defined *measurement* as a synthesis of change of position and subdivision (Piaget et al., 1960). More precisely, measurement for Piaget involved (a) understanding space or the length of an object as being partitionable, or as being able to be subdivided (subdivision), and (b) partitioning off a unit from an object and iterating that unit without overlap or empty intervals (change of position). Piaget and his colleagues argued that the coordination of these two notions along with the understanding that these continuous units form inclusions (i.e., the first length measured is included in the length that comprises two units, etc.) leads to a full understanding of measurement. The claim that conservation of length develops as a child learns to measure is implicit in this definition. *Conservation of length* means that as a child moves an object, the object's length does not change. Conservation of length is not equivalent to the concept of measurement but, rather, develops as the child learns to measure (Inhelder, Sinclair, & Bovet, 1974).

For Piaget, knowledge of measurement required that a child be able not only to apply the procedures and skills of measuring but also to carry out these activities without merely following specific, formalized measuring procedures. In other words, measurement for Piaget did not consist of simply performing the steps required to measure. Rather, children who could measure in his terms were almost immediately able to use the standard measure with insight, not by trial and error (Piaget et al., 1960). Consequently, a child who could use a ruler correctly to measure would not necessarily have an operational understanding of measurement according to Piaget and others' definition. Piaget and his colleagues' analyses have particular value because they focused on the mathematical *meanings* and *interpretations* rather than solely the methods of measuring associated with particular measurement understandings. These types of analyses are also consistent with current mathematical reform efforts. For example, the NCTM *Principles and Standards* document (2000) emphasizes that mathematics should be meaningful activity rather than a set of learned rules and procedures for calculating.

Piaget et al. (1960) categorized children's development of measurement conceptions into either three or four stages, depending on which aspect of measurement was being investigated. For example, the development of the conservation of length fell into three distinct stages with various substages. What follows is a brief, integrative description of some of the main stages in Piaget's account of children's development of measurement concepts. For a more detailed inspection of the various stages of development, see Piaget et al. (1960), Carpenter (1976), Copeland (1974), and Sinclair (1970).

At the initial stages of development, including various substages, children conserve neither length nor distance. Piaget used the term *distance* to refer to the empty space between two objects, such as two trees. He used the term *length* to denote the physical extension of an object, such as a stick. Children who are classified as being at early stages of development rely strictly on visual perception for judgments about length. For instance, Piaget asked several children in clinical interviews about two strips of equal length (see Figure 2.1). He placed them in direct alignment with each other, as shown in situation (1) in Figure 2.1, and asked the child which strip was longer. Once the child acknowledged equality of length, the interviewer moved the bottom strip a few centimeters, as shown in situation (2) in Figure 2.1, and again asked which one was longer.

Those children classified as in the early stages of development concluded that the bottom strip was now longer than the top strip because the bottom strip extended beyond the top one. Piaget concluded that these children were reasoning by relying

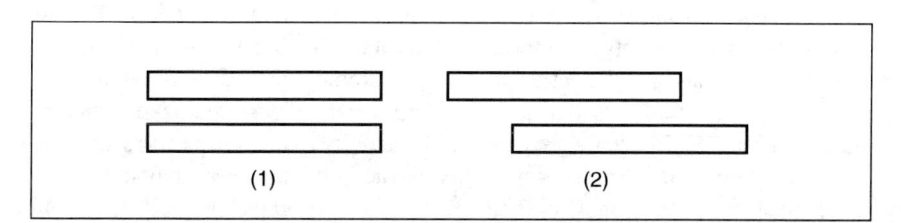

Figure 2.1. Interview task focusing on conservation of length.

on perceptual judgments centering on the position of the endpoints. He argued that operational measurement is not possible for children classified as in early stages, because space is not viewed as a common medium containing objects with well-defined spatial relations between them. According to Piaget, children with such an uncoordinated view of space do not understand that the distance from A to B is the same as the distance from B to A.

Piaget argued that children whom he classified as in the early stages were not able to reason transitively, an ability that involves, for instance, using a stick as an instrument to judge whether two immovable towers are the same size. Being able to reason transitively is essential for operational measurement. A child who can reason transitively can take a third or middle item (e.g., the stick) as a referent by which to compare the heights or lengths of other objects. During the early stages of development, however, when children do not conserve length, transitive reasoning is impossible because once children move the middle object (e.g., the stick), the length of that middle object, in their view, can change.

Piaget asserted that children classified as in the early stages do not have an understanding of either subdivision or displacement. This lack of understanding can be seen as children measure the length of an item with a smaller unit. Typically, these children either run the smaller unit alongside the length of the item without properly partitioning the length or they make iterations of unequal spaces, sometimes neglecting gaps or overlapping areas. Their thinking at these early stages is said to be intuitive and irreversible (i.e., they are unable to go backward; for instance, A to B is not the same as B to A).

Piaget identified transitional stages between these early stages and the final stages of operational measurement. Children in these transitional stages (about 6 to 7 years old) begin to reason transitively by using their body as a middle referent. Later, at about 7 to 8 years old, they can reason transitively with other objects and begin to conserve length. Piaget asserted, however, that children classified as in a transitional stage have not yet coordinated change of position and subdivision but possess each of the notions separately. These children understand that changing the position of an item does not alter quantity and that the whole is the sum of its parts, but they do not coordinate these two concepts.

Finally, the child attains the last stage, operational measurement, when he or she has synthesized displacement with subdivision (about 8 to 10 years old). Implicit in this acquisition for Piaget was that measuring consisted of a series of nested relations. In other words, as a unit is iterated, the distance covered by the first two units is nested in the distance covered by three units, and so on. Differentiating this last stage from the mere ability to perform the skills of conventional measuring is important. To be operational, the actions of the measurement process must be interiorized into conceptual acts that are an integral part of an organized structure (Carpenter, 1976). In other words, the child's activity of measuring correctly does not indicate operational measuring; rather, the activity must be accompanied by a conceptual reorganization. Children at this stage can compare units' lengths and discover that a small unit is a third or half of another. Similar to Piaget's perception, we characterize acting in a spatial environment as coordinating units with different measures, an ability that includes structuring parts of units in fractional units. In addition, the results of measuring should come to signify a series of nested distances (what we call *accumulation of distance*).

In Piaget's developmental stage theory, children pass from one stage to the next by making shifts in how they reorganize their previous experiences. Children proceed through these stages in a relatively fixed order. Piaget's developmental stage theory seems to characterize learning as a series of clear-cut cognitive reorganizations that signal that an individual has passed from one stage to the next. This characterization is individualistic in that Piaget explained learning as occurring solely in the mind of the child. Piaget would not deny that conceptual reorganizations often occur *as a consequence* of social interaction, but his analysis of children's learning to measure was cast *primarily* as an individual accomplishment. In contrast, current studies have demonstrated the social nature of learning (Lave, 1988; Rogoff, 1990; Saxe, 1991; Vygotsky, 1978). Piaget et al.'s (1960) analysis that children pass through a fixed set of stages did not account for social aspects of the learning process. The study that we report in this monograph can be viewed as drawing on Piaget and his colleagues' description of what counts as a deep understanding of measurement (e.g., operational measurement as subdivision and displacement and inclusive), yet it recasts cognition as occurring in situ. In other words, we find Piaget and others' psychological analysis to be extremely valuable for determining the instructional intent of the measurement design experiment but insufficient for analyzing the learning of students as they participate in the social context of a classroom community. Therefore, we draw on Piaget's conceptual distinctions regarding learning to measure to make sense of children's learning in social context.

Reactions to Piaget

Researchers, sparked by Piaget's analysis of children's conceptions of measurement, focused on two different sets of issues. One interpretation of Piaget's developmental model was that certain measurement concepts developed in fixed order. For example, some researchers argued that the conservation of length developed before transitive reasoning. Also, some researchers interpreted the ages at which children construct particular measurement concepts as being relatively fixed. Researchers commonly tested the order in which measurement concepts developed or debated the ages at which children reasoned transitively. As a consequence, many researchers sought to substantiate or discredit Piaget's earlier claims. Along this same line, more recent studies have used Piaget's epistemology to detail the conceptual schemas that children construct as they learn to measure. Thus, I call the first collection of measurement literature "cognitive analyses."

A second set of investigations based on Piaget's analyses focused on increasing the acquisition rate of certain measurement concepts (Beilin, 1971). For example, researchers attempted to train children to conserve length by using a variety of training techniques. The primary focus of the "training studies" was to increase the rate at which children learn the conservation and transitivity of length using various training techniques. Therefore, many different experimental methods were being used.

In the following sections, I synthesize the majority of the studies based on Piaget's analyses. Because the researchers made a variety of different and sometimes conflicting theoretical assumptions, however, definitive conclusions about their findings are very difficult to draw. Thus, I describe briefly the most common theoretical perspectives that guided the research. My intent is not to present an exhaustive account of the research but rather to provide an overview of the topics that were studied and debated.

Cognitive analyses. In this first set of studies, what I have termed "cognitive analyses," many researchers tested the order of Piaget's measurement concepts by using different experimental methods and tasks (e.g., Carpenter, 1975; Gelman, 1969; McManis, 1969; Murray, 1965, 1968; Shantz and Smock, 1966). For instance, Shantz and Smock conducted a study investigating whether the development of the conservation of length occurs before the development of a spatial coordinate system. Their findings supported Piaget and Inhelder's (1956) theory that the development of the conservation of length, which occurs around age 7, precipitates the development of a spatial coordinate system, which occurs at about age 9. Generally, earlier research on the age and order of the development of measurement concepts supported Piaget and Inhelder's claims. In more recent studies, however, researchers argued that conservation and transitivity do not necessarily need to be constructed first after all. In fact, they contend that children can construct the inverse relation between number and unit-size before conservation (Clements, 1999; Hiebert, 1981; Petitto, 1990) but that conservation is an important notion that leads to operational measurement (Petitto, 1990).

A second body of research focused on the age at which students learn transitivity. Transitivity studies were the subject of so many methodological variations that researchers argued that the ages at which transitivity developed in children ran from as early as 4 years to as late as 8 years old. As a result, a debate ensued between Braine (1964) and Smedslund (1963, 1965) concerning assessment techniques and the need to define more exactly what was meant by *evidence of transitivity*. Both Braine and Smedslund accounted for learning using Piaget's developmental model, but they disagreed on the ages at which transitivity occurred; Braine suggested that transitivity occurred at 4 to 5 years old, and Smedslund suggested that it occurred at 7 to 8 years old. A more recent study by Kamii (1997) confirmed Piaget and Inhelder's (1956) and Smedslund's findings. She also found that transitivity must be developed before unit iteration. As to Piaget's earlier claims that transitivity and conservation develop simultaneously, some researchers performed experiments that negated this claim (Brainerd, 1974; Lovell & Ogilvie, 1961; McManis, 1969; Smedslund, 1961; Steffe & Carey, 1972).

With regard to constructing a full, operational understanding of measurement, Lovell, Healey, and Rowland (1962) generally substantiated Piaget and Inhelder's

(1956) stage theory categorizations. Also, the ages at which full measurement develops in children were found to be close to Piaget's estimate of 9 years old. Kamii (1997) obtained similar results in her study, suggesting that full measurement understanding develops around 9 to 10 years old. Lehrer, Jenkins, & Osana (1998) substantiated this age range when they found that children below fourth grade incorrectly measured the length of an object using two different-sized units. Lehrer and his colleagues concluded that these students had not constructed the relation between unit of measure and the attribute being measured (see also Barrett and Clements, 1996; Clements, Battista, & Sarama, 1998).

Although Piaget and other researchers often disagreed, all these studies share one assumption. In these analyses, learning is primarily analyzed as an individualistic achievement. Learning consists of cognitive reorganizations that occur in the mind of the child, sometimes *as a consequence* of social interaction or tool use. In other words, the researchers realized that the process of learning could have social aspects, but analyses were cast solely in terms of an individual's cognitive development.

Training studies. The second body of literature, training studies, came about as a reaction to Piaget's naturalistic slant on learning. Piaget developed a general theory of conceptual development for transitions between stages, but most of this research focused on children's performing tasks presented in clinical interviews. Very little empirical evidence supported his theories about the process and ways of supporting conceptual development. Consequently, training studies centered on the acquisition of cognitive structures by considering the effect of different types of training procedures.

Piaget described training studies as an attempt to rush development (Rogoff, 1990), and he is generally interpreted as doubting that training has an effect on the development of children's conceptions. Nevertheless, many researchers based their training techniques on the basic components of Piaget's equilibrium model. Equilibration is best described as a dynamic, self-organized balance between assimilation and accommodation. Assimilation is the process of organizing new experiences in terms of prior understanding. When new experiences cannot be adequately organized in this way or when they give rise to contradictions, an accommodation typically occurs. An accommodation involves reflective, integrative processes that create a new understanding to restore equilibrium. Piaget conjectured that disequilibrium is brought about by experiences that generate cognitive perturbations (von Glasersfeld, 1995).

Many of the training programs used aspects of the equilibrium model as the assumptions behind their training methods. Researchers tried to induce cognitive conflict in children to bring about disequilibrium (Bailey, 1974; Brainerd, 1974; Murray, 1968; Overbeck & Schwartz, 1970; Smedslund, 1961). Other researchers argued that reversibility might play an important role in decreasing the age at which conservation of length developed in children (Brison, 1966; Murray, 1968; Smith, 1968). For instance, one of the experimental treatments in Smith's (1968) training program involved asking children to make judgments about the length of an object after adding and subtracting pieces from the end of it. In this way, the child was "shown" that reversing operations could help determine equality of objects. Like the cognitive analyses, training studies also focus primarily on individual cognitive growth and hence do not include descriptions of any social aspects or tool use involved in the training sessions.

Although these training studies were based on aspects of Piaget's equilibrium model to determine training methods, their training goals seem to have differed from Piaget's goals. In my view, Piaget was concerned with identifying children's understanding of, and their constructed meanings for, conservation. In contrast, analyses of the training studies focused only on whether children could conserve length by the end of the training sessions. Their primary concern was on children's acquisition of the conservation of length, and this attainment was determined by correct answers to conservation tasks. In other words, evidence that a child conserved length was determined by his or her performance on conservation tasks, not by detailed analyses of his or her thinking. One exception to this technique is a study conducted by Inhelder et al. (1974). This study differed in that preliminary experiments were conducted to understand children's conceptions of the relationship between number and length and to identify stages of development pertaining to this relationship. After the stages were identified, training sessions were conducted and several of the children's responses during the session were analyzed to identify the psychological processes involved in coordinating discrete units with continuous length. The main product of these training sessions consisted of psychological analyses of children's solutions, not just scores on tests. On the one hand, this kind of study reflects Piaget's theory of learning more than the other training studies. On the other hand, such analyses were still cast in terms of cognitive reorganizations devoid of social context.

Another set of training studies focused on whether children transferred their measurement concepts to other task situations (or domains). For example, many researchers investigated whether children who were trained to conserve length would transfer those skills to conserve weight or area (Beilin, 1965; Brainerd, 1974; Brison, 1966; Gruen, 1965; Tomic, Kingma, & Tenvergert, 1993). Training success was determined by comparing scores on pretests and posttests instead of attempting to explain children's meaningful activity on a variety of interview tasks. Generally, researchers found that training for the conservation of length using various techniques transferred to other domains (cf. Beilin, 1965; Brainerd, 1974). For a more in-depth review of the literature concerning transfer of training, see Osborne (1976).

A final group of training studies was conducted from an information-processing perspective. Studies had been conducted in the 1970s by researchers trying to identify the information-processing demands of tasks that dealt with the conservation of length (Baylor, Gascon, Lemoyne, & Pothier, 1973; Hiebert, 1981; Klahr and Wallace, 1970). The computer was the dominant metaphor with which to describe children's mathematical learning, as in the following:

It was because the activities of the computer itself seemed in some ways akin to cognitive processes. Computers accept information, manipulate symbols, store items in "memory" and retrieve them again, classify inputs, recognize patterns and so on. (Gardner, 1987, p. 119)

Human thought was conceptualized as an information-processing system, and the goal was to develop computational models whose input-output relations matched those of children's observed performance (Cobb, 1990). In some cases, researchers actually attempted to write computer programs that matched their observations of children's strategies (e.g., Baylor et al., 1973). According to these researchers, many mathematics tasks require the ability to process several pieces of information simultaneously, and students with low short-term-memory capacities were assumed to perform less well than children with high capacities. Therefore, researchers took into account the information-processing requirements of a task as they analyzed students' solutions to problems. Hiebert (1981) and Baylor et al. (1973) studied the relationship between information-processing capacities and children's ability to learn concepts of linear measurement and found that the information-processing capacity had no detectable influences. Nonetheless, Hiebert argued along with Baylor and his colleagues that the usefulness of this perspective in educational settings requires more research (see also Inhelder, 1972; Klahr & Wallace, 1970). A more recent study outlined the sequence in which measurement concepts should be taught and based their ordering on the information-processing demand of the tasks (Boulton-Lewis, 1987). This approach, as other computational models of the information-processing and training theories do more generally, focuses primarily on cognitive aspects of training. As such, the approach influenced the work of researchers attempting to learn what types of interventions work, but it did not generate insights regarding what social aspects or tools affected the observed outcomes.

Summary of Measurement Investigations

Throughout this review, I have alluded to several common themes among the various studies. In this section, I summarize the findings from the literature to describe three main issues that appear to run throughout the literature regarding Piagetian development: (1) learning was analyzed primarily from an individual perspective, (2) most studies involved documenting the criteria for the success of training studies, and (3) very few studies emphasized the role of tools in supporting students' development. Of these three main issues, I identify the theoretical themes that run throughout the monograph.

The first issue running throughout the literature involved characterizing learning as an individual accomplishment. In most cases, social aspects were seen as serving as catalysts for cognitive reorganizations, but their role as such was not explicitly described. For example, consider the debate between Smedslund (1961) and Brainerd (1974) regarding the age at which students learn transitivity. Smedslund interpreted Piaget as giving very little importance to social processes (i.e.,

Smedslund argued that Piaget would claim that positive feedback from the interviewer was not the cause for conceptual reorganizations, because they were social processes; conceptual reorganizations, in Smedslund's interpretation, were caused by the individual). Brainerd countered that Piaget would argue that positive feedback *could* cause cognitive perturbations. Both interpretations explained learning in terms of cognitive reorganizations (at times caused by social events) and placed little emphasis on learning as a social process. At the beginning of this chapter, I pointed out that current theories emphasize placing more than a secondary role on the social processes involved in learning. As a consequence of these current theories, further investigations of children's development of measurement conceptions might focus on students' activity in social context. Thus, one theoretical theme that follows naturally out of this literature review concerns paying explicit attention to learning as it occurs in the *social context of the classroom*.

A second issue illuminated in the literature review concerns the criteria for success of the training studies. A training technique was judged successful if posttest scores were significantly higher than pretest scores. The means of supporting children's development, however, were not described in the analysis of training sessions. In other words, if training techniques did, in fact, increase the likelihood that students could conserve length, then training was considered a success. Analyses of students' learning during these training sessions were not tied to the ways the researchers engaged with them that might have led students to make their conceptual reorganizations during training sessions. Current instructional design theories call for analyses of students' learning to be tied to our (i.e., teachers', trainers', or researchers') role in supporting that learning (Cobb, Stephan, McClain, and Gravemeijer, 2001). In other words, the goals of a researcher are not only to improve students' learning but also both to reflect on the process by which students developed and to locate themselves as designers in that process. Thus, analyses of students' learning in social context can feed back to inform designers of the ways in which instructional sequences (or training techniques) can be revised to better support students' learning in the social context of the classroom setting. The second theme, therefore, that runs throughout this monograph focuses on ways of *proactively supporting students' development from the designers' perspective*.

A third issue running throughout the literature review centers on the role of tools in supporting or constraining students' mathematical development. With the exception of a few studies (e.g., Barrett & Clements, 1996; Clements et al., 1998; Nunes, Light, & Mason, 1993; Piaget et al., 1960), very few of the prior studies on measurement focused on the role that measuring tools play in supporting students' mathematical development. When tools were considered as part of the child's development, the nature of these tool analyses was individualistic as well. For example, Piaget and his colleagues ascribed a child's assimilation or accommodation of a tool to previous measuring experiences. Again, it was as if the tool served as a *catalyst* for cognitive reorganizations but was not an integral aspect of what was learned. In contrast, current theories suggest that further measurement analyses consider tools as an intimate part of students' mathematical development. Currently in the field of education, many theorists have emphasized that learning does not occur apart from tool use (cf. Cobb, 1997; Kaput, 1994; Meira, 1995, 1998; Nemirovsky & Monk, 2000; Pea, 1993; van Oers, 1996; Vygotsky, 1978). Some researchers go further by arguing for an extremely strong relationship between tool use and learning. They contend that symbol creation and meaning making are not separate endeavors but occur simultaneously (Meira, 1995, 1998). Because measuring clearly involves using measurement tools, a third and final theme concerns supporting students' mathematical development as they reason with *cultural tools*.

RECONCEPTUALIZING MEASUREMENT INVESTIGATIONS

Theoretical growth in the field of mathematics education allows us, as researchers, to look back on the tremendous work of our predecessors and build on their measurement findings to learn more about supporting students' development of measurement understandings from new perspectives. Current theories on social and cultural processes and new methods of proactively supporting students' mathematical development provide new lenses with which to view old problems. In this section I develop our theoretical position on each of the three literature review themes discussed above: (1) social aspects of learning, (2) proactively supporting students' learning, and (3) cultural tools.

Social Aspects of Learning

All the studies discussed in the literature review involved a primarily individualistic view of learning. An increasing number of researchers, however, have adopted the premise that learning occurs in a socially situated context and that these social and cultural aspects greatly influence the way individuals construct their understandings (cf. Cobb & Bauersfeld, 1995; Cole, 1996; Lave, 1988; Rogoff, 1990; Saxe, 1991; Vygotsky, 1978; Wenger, 1998).

One of the consequences of adopting such a position is that whereas most researchers acknowledge that learning has both social and individual qualities, the emphasis they place on each perspective may differ significantly. For example, whereas Piaget and his colleagues (Piaget et al., 1960; Piaget and Inhelder, 1956) cast analyses in terms of cognitive reorganizations, with social aspects serving merely as a secondary catalyst, some theorists take a highly contrasting viewpoint (cf. Lerman, 1996). For example, some socioculturalists have maintained that learning is first and foremost a social endeavor, with individual processes considered to be secondary. This argument is often supported by the following commonly cited quote by Vygotsky:

> Any function of the child's development appears twice, or on two planes. *First* it appears on the social plane and then on the psychological plane. *First* it appears between people as an interpsychological category and then within the child as an intrapsychological category. (Vygotsky, 1978, p. 63) (italics added for emphasis)

The authors of this monograph attempt to transcend these two extreme positions and view learning as a process that involves both individual and social aspects with *neither taking primacy over the other*. This way of viewing learning is a version of social constructivism called the *emergent perspective*. In this view, the relationship between social and individual processes is extremely strong, in that the two cannot be separated—the existence of one depends on the existence of the other. The result of this assumption is that researchers assuming the emergent perspective attempt to account for individuals' mathematical development as they participate in the social and cultural practices of the classroom community (Cobb, 2000; Yackel and Cobb, 1995). On the one hand, individual students' development is analyzed in terms of their participation in, and contribution to, the emerging, communal mathematical practices. On the other hand, mathematical practices are the taken-as-shared ways a community comes to reason and argue mathematically (Cobb et al., 2001; Stephan, 1998) and often involve aspects of symbolizing. (Chapters 3 and 5 of this monograph further elaborate the meaning of the construct of a mathematical practice.) Thus, we can say that the relationship between individual and social process is strong because although practices constitute the taken-as-shared learning of a community, we take it as equally important to acknowledge that students are seen to contribute to the evolution of the classroom mathematical practices as they reorganize their mathematical activity. In other words, mathematical practices and individual students' conceptual development are seen as reflexively related. Whereas practices are established and evolve as students participate in, and contribute to, them, these acts of participation constitute, for us, acts by individual students of reasoning and possible conceptual reorganizations.

Although a strong dichotomy seems to exist between situated and individual perspectives at the extremes (consider, e.g., the recent debate between Lerman (2000) and Steffe & Thompson (2000)), the emergent perspective attempts to coordinate the two positions (cf. Confrey, 1995). The classroom teaching experiment reported in this monograph is an example of an investigation that places students' understandings of measurement in the local mathematical practices of their classroom community.

Despite different perspectives regarding the role of social and cultural processes in learning, most researchers in this area agree that students' development cannot be adequately explained in cognitive terms alone; social and cultural processes must be acknowledged when explaining mathematical development. These social theories, of course, do not discount psychological analyses conducted in interviews. Theorists from both sociocultural and emergent perspectives, however, would argue that traditional psychological analyses of interview situations characterize students' conceptual understanding independently of situation and purpose (Cobb & Yackel, 1996) and thus must be complemented by social analyses to gain a more comprehensive view of the learning process as it evolves.

Michelle Stephan

Proactively Supporting Students' Learning

A second theme that was illuminated by the prior literature concerned the role that designers and researchers play in supporting students' mathematical development. The prior training studies showed that students' scores on pretests and posttests determined the success of the training techniques and thus whether students constructed the targeted conceptual structures. The analyses of students' progress, therefore, focused on neither the process of their development nor the researchers' means of supporting conceptual reorganizations.

In contrast, we have been in the process of refining an approach to classroom-based investigations that both situates analyses of students' learning in social context and ties these analyses to our means of supporting development. This type of research, called Design Research (Brown, 1992; Cobb et al., 2001; Collins, 1999), involves a reflexive relationship between classroom-based research and instructional design (see Figure 2.2).

```
Instructional design                Classroom-based research
 guided by domain-specific            guided by interpretive
 instructional theory                 framework
```

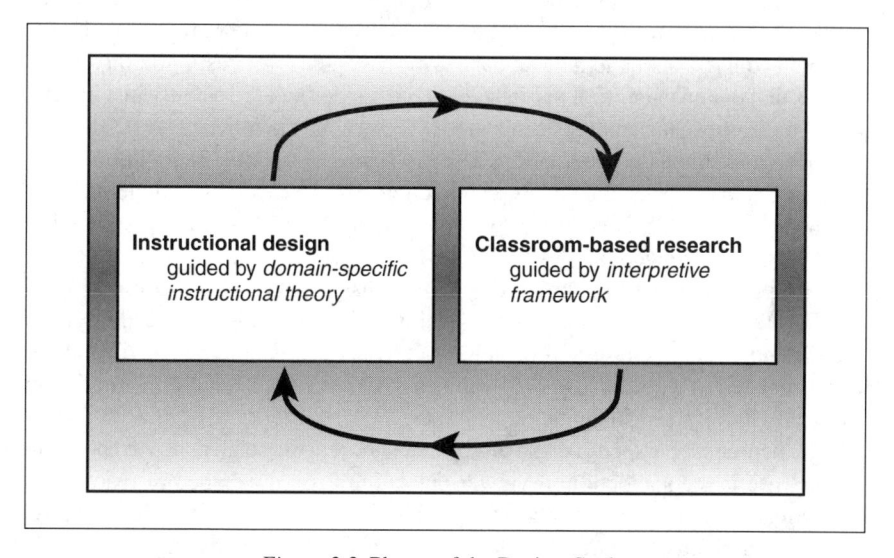

Figure 2.2. Phases of the Design Cycle.

Gravemeijer (1994) has written extensively about the process of design. In brief, he explains that designers engage in an anticipatory thought experiment in which they anticipate how the taken-as-shared mathematical learning and discourse might evolve as proposed instructional activities are enacted. This conjecturing amounts to developing a potential taken-as-shared learning route and the means by which designers might support such mathematical activity. These conjectures are provisional and are tested and modified in the course of classroom-based experimentation. As Cobb and his colleagues (2001) note, the process of testing and refining is guided by classroom-based research, the second aspect of the Design Cycle. These analyses feed back to inform the designer of both the role of the instructional activities in supporting the learning of the community and possible changes in the instructional sequence. This way of proceeding differs from previous studies conducted on measurement in that a more detailed account of children's learning as it occurred in social context is documented. Further, the success of the measurement training studies was determined by pretest and posttest scores. Thus the process of supporting students' learning was left unexamined. In contrast, Design Research is one viable model that accounts for both the process of students' measurement conceptions and the ways of supporting that development.

Cultural Tools

For the most part, the measurement analyses discussed throughout this chapter did not take into account students' activity with cultural tools, which, for our purposes include physical materials, tables, pictures, computer graphs and icons, and both conventional and nonstandard symbol systems. Interestingly, those four studies that did consider tool use in students' development did so only in a limited way. For example, the analyses conducted by Piaget and his colleagues (1960) characterized individual development as a consequence of assimilating cultural tools with prior experiences. The analyses conducted by Barrett & Clements (1996), Clements et al. (1998), and Nunes et al. (1993) all focused on the ways in which tools facilitated conceptual development and served as catalysts for developing thought processes. Current theories contend that conceptual development and symbol creation are reflexively related (Meira, 1995, 1998; van Oers, 2000). These theories hold that symbols and meanings are constructed simultaneously and are consistent with Vygotsky's (1978) view emphasizing that semiotic mediation is integrally involved in students' development.

One theoretical perspective that has been developing a socially situated position on tool use is the sociocultural view. Socioculturalism developed in reaction to mainstream psychology's attention to individual activity. Socioculturalists emphasize the complexity of learning within a social environment and suggest that tools have a direct influence on how individuals learn. The notion of distributed intelligences, which draws heavily on sociocultural theories, suggests that students may use cultural tools as scaffolds to their internalization process. Pea (1993) describes tool use as a reorganizer, rather than amplifier, of understanding. Socioculturalists characterize tool use as a process of internalizing a cultural tool so that it becomes a tool for thinking (Davydov & Radzikhovskii, 1985) or of appropriating it to one's own activity (Newman, Griffin, & Cole, 1989). These and other sociocultural theorists often argue that tools are the primary vehicles for enculturation into the cultural practices of the wider community because tools are carriers of meanings from one generation to the next (van Oers, 1996). In this view, students are said to inherit the cultural meanings involved in using a particular tool. From the sociocultural perspective, the tools and their associated, culturally approved meanings are objects to be internalized by the individual. Nunes et al. (1993) took this approach when analyzing their students' development of measuring conceptions.

In the emergent perspective, tools are also viewed as a constituent part of individuals' activities. Further, emergent theorists would argue that the meanings associated with tools are created and evolve as students reason with physical materials, pictures, and so forth (Cobb and Yackel, 1996). This perspective is in marked contrast with sociocultural views in which *established* cultural meanings are said to be internalized by the individual. Along with Meira (1995), emergent theorists contend that the meanings involved in acting with a particular tool evolve as students are engaged in goal-directed activity with the tool. From the emergent perspective, the idea that learning occurs as students act with cultural tools is based on the assumption that the symbolic and conceptual meanings that evolved from students' activity with tools do not develop apart from the mathematical practices in which students participate. The teacher's role, then, is to initiate and guide the development of individuals' mathematical activity—in particular, tool activity—as well as the mathematical practices so that these practices become more compatible with those of the wider society (Cobb, 1994).

Although ways of analyzing and describing tool use are still being investigated, attending to the role of semiotic mediation in measurement clearly is of great importance. Cultural tools were used to a large extent in training students to conserve length, to reason transitively, and to learn to measure. The question remains as to how students' activity with these tools changed the way they came to view measuring. With this question in mind, we hope to build on the important work of those researchers who have already begun to highlight tool use in students' measurement development (e.g., Barrett & Clements, 1996; Clements et al., 1998; Nunes et al., 1993).

CONCLUSION

In this chapter, I used the prior literature as background to develop three theoretical themes that lay the foundation for this monograph. The first of these themes concerned outlining the theoretical perspective for documenting students' mathematical (i.e., measuring) activity as they participate in the social context of their classroom community. The second theme focused on issues of instructional design and ways of supporting students' measurement development. The third theme dealt with the evolution of the collective meanings associated with tools used to support students' measuring activity. Each of these themes reflects an orientation toward learning as both a social and individual process. The view I have portrayed is one in which social and individual processes are reflexively related. In other words, social and individual processes coexist, and learning in the context of classroom-based instruction cannot be adequately explained in terms of either one or the other. Also, an argument has been made that individual as well as collective activity cannot be separated from activity with cultural tools. Therefore, in specifying the mathematical practices and individual mathematical development within these practices, activity with cultural tools becomes a central focus. In the next chapter of the monograph, we expand more generally on our instructional design theory while integrating the other two research themes into the discussion.

REFERENCES

Bailey, T. (1974). Linear measurement in the elementary school. *The Arithmetic Teacher, 21,* 520–525.

Barrett, J. E., & Clements, D. H. (1996). *Representing, connecting and restructuring knowledge: A microgenetic analysis of a child's learning in an open-ended task involving perimeter, paths, and polygons.* Paper presented at the Eighteenth Annual Meeting of the North American Chapter of the International Group for the Psychology of Mathematics Education, Panama City, Florida.

Baylor, W., Gascon, J., Lemoyne, G., & Pothier, N. (1973). An information processing model of some seriation tasks. *Canadian Psychologist, 14,* 167–196.

Beilin, H. (1965). Learning and operational convergence in logical thought development. *Journal of Experimental Child Psychology, 2,* 317–339.

Beilin, H. (1971). The training and acquisition of logical operations. In M. Rosskopf, L. Steffe, & S. Taback (Eds.), *Piagetian cognitive-development research and mathematical education* (pp. 81–124). Washington, DC: NCTM.

Boulton-Lewis, G. (1987). Recent cognitive theories applied to sequential length measuring knowledge in young children. *British Journal of Educational Psychology, 57,* 330–342.

Braine, M. (1964). Development of a grasp of transitivity of length: A reply to Smedslund. *Child Development, 35,* 799–810.

Brainerd, C. (1974). Training and transfer of transitivity, conservation, and class inclusion of length. *Child Development, 45,* 324–334.

Brison, D. (1966). The acceleration of conservation of substance. *Journal of Genetic Psychology, 109,* 311–332.

Brown, A. L. (1992). Design experiments: Theoretical and methodological challenges in creating complex interventions in classroom settings. *Journal of the Learning Sciences, 2,* 141–178.

Carpenter, T. (1975). Measurement concepts of first- and second-grade students. *Journal for Research in Mathematics Education, 6,* 3–13.

Carpenter, T. (1976). Analysis and synthesis of existing research on measurement. In R. Lesh & D. Bradbard (Eds.), *Number and measurement.* [Papers from a research workshop]. (ERIC Document Reproduction Service No. ED 120027)

Clements, D. (1999). Teaching length measurement: Research challenges. *School Science and Mathematics, 99*(1), 5–11.

Clements, D., Battista, M., & Sarama, J. (1998). Development of geometric and measurement ideas. In R. Lehrer and D. Chazan (Eds.), *Designing learning environments for developing understanding of geometry and space* (pp. 201–226). Hillsdale, NJ: Erlbaum.

Cobb, P. (1990). A constructivist perspective on information-processing theories of mathematical activity. *International Journal of Educational Research, 14,* 67–92.

Cobb, P. (1994). *Theories of mathematical learning and constructivism: A personal view.* Paper presented at the Symposium on Trends and Perspectives in Mathematics Education, Institute for Mathematics, University of Klagenfurt, Austria.

Cobb, P. (1997). Learning from distributed theories of intelligence. In E. Pehkonen (Ed.), *Proceedings of the Twenty-First International Conference for the Psychology of Mathematics Education* (Vol. 2, pp. 169–176). Finland: University of Helsinki.

Cobb, P. (2000). Conducting teaching experiments in collaboration with teachers. In A. E. Kelly & R. A. Lesh (Eds.), *Handbook of research design in mathematics and science education* (pp. 307–334). Mahwah, NJ: Erlbaum.

Cobb, P., & Bauersfeld, H. (1995). The coordination of psychological and sociological perspectives in mathematics education. In P. Cobb & H. Bauersfeld (Eds.), *The emergence of mathematical meaning: Interaction in classroom cultures.* Hillsdale, NJ: Erlbaum.

Cobb, P., Stephan, M., McClain, K., & Gravemeijer, K. (2001). Participating in mathematical practices. *Journal of the Learning Sciences, 10*(1, 2), 113–163.

Cobb, P. & Yackel, E. (1996). Constructivist, emergent, and sociocultural perspectives in the context of developmental research. *Educational Psychologist, 31*, 175–190.

Cole, M. (1996). *Cultural psychology*. Cambridge, MA: Belknap Press of Harvard University Press.

Collins, A. (1999). The changing infrastructure of educational research. In E. C. Langemann & L. S. Shulman (Eds.), *Issues in education research* (pp. 289–298). San Francisco: Jossey Bass.

Confrey, J. (1995). How compatible are radical constructivism, socioculturalist approaches, and social constructivism? In L. Steffe & J. Gale (Eds.), *Constructivism in education* (pp. 185–225). Hillsdale, NJ: Erlbaum.

Copeland, R. (1974). *How children learn mathematics: Teaching implications of Piaget's research* (2nd ed.). New York: Macmillan.

Davydov, V., & Radzikhovskii, L. (1985). Vygotsky's theory and the activity-oriented approach in psychology. In J. Wertsch (Ed.), *Culture, communication, and cognition: Vygotskian perspectives* (pp. 35–65). New York: Cambridge University Press.

Gardner, H. (1987). *The mind's new science: A history of the cognitive revolution*. New York: Basic Books.

Gelman, R., (1969). Conservation acquisition: A problem of learning to attend to relevant attributes. *Journal of Experimental Child Psychology, 7*, 167–187.

Gravemeijer, K. (1994). *Developing realistic mathematics education*. Utrecht, Netherlands: CD-β Press.

Gruen, G. (1965). Experiences affecting the development of number conservation in children. *Child Development, 36*, 963–979.

Hiebert, J. (1981). Cognitive development and learning linear measurement. *Journal for Research in Mathematics Education, 12*(3), 197–211.

Inhelder, B. (1972). Information processing tendencies in recent experiments in cognitive learning—empirical studies. In S. Farnham-Diggory (Ed.), *Information processing in children* (pp. 103–114). New York: Academic Press.

Inhelder, B., Sinclair, H., & Bovet, M. (1974). *Learning and the development of cognition*. Cambridge, MA: Harvard University Press.

Kamii, C. (1997). Measurement of length: The need for a better approach to teaching. *School Science and Mathematics, 97*(3), 116–121.

Kaput, J. J. (1994). The representational roles of technology in connecting mathematics with authentic experience. In R. Biehler, R. W. Scholz, R. Strasser, & B. Winkelmann (Eds.), *Didactics of mathematics as a scientific discipline* (pp. 379–397). Dordrecht, Netherlands: Kluwer.

Klahr, D., & Wallace, J. (1970). An information processing analysis of some Piagetian experimental tasks. *Cognitive Psychology, 1*, 358–387.

Lave, J. (1988). *Cognition in practice: Mind, mathematics, and culture in everyday life*. Cambridge, England: Cambridge University Press.

Lehrer, R., Jenkins, M., & Osana, H. (1998). Longitudinal study of children's reasoning about space and geometry. In R. Lehrer & D. Chazan (Eds.), *Designing learning environments for developing understanding of geometry and space* (pp. 137–167). Hillsdale, NJ: Erlbaum.

Lerman, S. (1996). Intersubjectivity in mathematics learning: A challenge to the radical constructivist paradigm? *Journal for Research in Mathematics Education, 27*(2), 133–168.

Lerman, S. (2000). A case of interpretations of "social": A response to Steffe and Thompson. *Journal for Research in Mathematics Education, 31*(2), 210–227.

Lovell, K., Healey, D., & Rowland, A. (1962). Growth of some geometrical concepts. *Child Development, 33*, 751–767.

Lovell, K., & Ogilvie, E. (1961). A study of the conservation of weight in the junior school child. *British Journal of Educational Psychology, 31*, 138–144.

McManis, D. (1969). Conservation and transitivity of weight and length by normals and retardates. *Developmental Psychology, 1*, 373–382.

Meira, L. (1995). The microevolution of mathematical representations in children's activities. *Cognition and Instruction, 13*(2), 269–313.

Meira, L. (1998). Making sense of instructional devices: The emergence of transparency in mathematical activity. *Journal for Research in Mathematics Education, 29*(2), 121–142.

Murray, F. (1965). Conservation of illusion distorted lengths and areas by primary school children. *Journal of Educational Psychology, 56*, 62–66.

Murray, F. (1968). Cognitive conflict and reversibility training in the acquisition of length conservation. *Journal of Education Psychology, 59*, 82–87.

National Council of Teachers of Mathematics (NCTM) (2000). *Principles and Standards for School Mathematics*. Reston, VA: National Council of Teachers of Mathematics.

National Council of Teachers of Mathematics (NCTM). (2003). *Learning and Teaching Measurement* (2003 Yearbook). Douglas H. Clements (Ed.). Reston, VA: NCTM.

Nemirovsky, R. C., & Monk, S. (2000). If you look at it the other way.... In P. Cobb, E. Yackel, & K. McClain (Eds.), *Symbolizing, communicating. and mathematizing: Perspectives on discourse, tools, and instructional design* (pp. 177–221). Mahwah, NJ: Erlbaum.

Newman, D., Griffin, P., & Cole, M. (1989). *The construction zone: Working for cognitive change in school*. Cambridge, MA: Cambridge University Press.

Nunes, T., Light, P., & Mason, J. (1993). Tools for thought: The measurement of length and area. *Learning and Instruction, 3*, 39–54.

Osborne, A. (1976). The mathematical and psychological foundations of measure. In R. Lesh & D. Bradbard (Eds.), *Number and measurement*. [Papers from a research workshop]. (ERIC Document Reproduction Service No. ED 120027).

Overbeck, C., & Schwartz, M. (1970). Training in conservation of weight. *Journal of Experimental Child Psychology, 9*, 253–264.

Pea, R. (1993). Practices of distributed intelligence and designs for education. In G. Salomon (Ed.), *Distributed cognitions* (pp. 47–87). New York: Cambridge University Press.

Petitto, A. (1990). Development of number line and measurement concepts. *Cognition and Instruction, 7*(1), 55–78.

Piaget, J., & Inhelder, B. (1956). *The child's conception of space*. London: Routledge & Kegan Paul.

Piaget, J., Inhelder, B., & Szeminska, A. (1960). *The child's conception of geometry*. New York: Basic Books.

Rogoff, B. (1990). *Apprenticeship in thinking: Cognitive development in social context*. Oxford, England: Oxford University Press.

Saxe, G. (1991). *Culture and cognitive development: Studies in mathematical understanding*. Hillsdale, NJ: Erlbaum.

Shantz, C., & Smock, C. (1966). Development of distance conservation and the spatial coordinate system. *Child Development, 37*, 943–948.

Sinclair, H. (1970). Number and Measurement. In M. Rosskopf, L. Steffe, & S. Taback (Eds.), *Piagetian cognitive-development research and mathematical education* (pp. 135–149). Washington, DC: National Council of Teachers of Mathematics.

Smedslund, J. (1961). The acquisition of substance and weight in children: II External reinforcement of conservation of weight and the operation of addition and subtraction. *Scandinavian Journal of Psychology, 2*, 71–84.

Smedslund, J. (1963). Development of concrete transitivity of length in children. *Child Development, 34*, 389–405.

Smedslund, J. (1965). The development of transitivity of length: A comment on Braine's reply. *Child Development, 36*, 577–580.

Smith, I. (1968). The effects of training procedures upon the acquisition of conservation of weight. *Child Development, 39*, 515–526.

Steffe, L., & Carey, R. (1972). Equivalence and order relations as interrelated by four- and five-year-old children. *Journal for Research in Mathematics Education, 3*, 77–88.

Steffe, L., & Thompson, P. (2000). Interaction or intersubjectivity? A reply to Lerman. *Journal for Research in Mathematics Education, 31*(2), 191–209.

Stephan, M. L. (1998). *Supporting the development of one first-grade classroom's conceptions of measurement: Analyzing students' learning in social context.* Unpublished doctoral dissertation, Vanderbilt University.

Tomic, W., Kingma, J., & Tenvergert, E. (1993). Training in measurement. *Journal of Educational Research, 86*(6), 340–348.

van Oers, B. (1996). Learning mathematics as meaningful activity. In P. Nesher, L. Steffe, P. Cobb, G. Goldin, & B. Greer (Eds.), *Theories of mathematical learning* (pp. 91–113). Hillsdale, NJ: Erlbaum.

van Oers, B. (2000). The appropriation of mathematical symbols: A psychosemiotic approach to mathematical learning. In E. Y. P. Cobb & K. McClain (Eds.), *Symbolizing and communicating in mathematics classrooms: Perspectives on discourse, tools, and instructional design* (pp. 133–176). Mahwah, NJ: Erlbaum.

von Glasersfeld, E. (1995). *Radical constructivism: A way of knowing and learning.* Bristol, PA: The Falmer Press.

Vygotsky, L. (1978). *Mind in society.* Cambridge, MA: Harvard University Press

Wenger, E. (1998). *Communities of practice.* New York: Cambridge University Press.

Yackel, E., & Cobb, P. (1995). Classroom sociomathematical norms and intellectual autonomy. In L. Meira & D. Carraher (Eds.), *Proceedings of the 19th International Conference for the Psychology of Mathematics Education* (vol. 3, pp. 264–272). Recife, Brazil: Program Committee of the 19th PME Conference.

Chapter 3

The Methodological Approach to Classroom-Based Research

Michelle Stephan
Purdue University Calumet
Paul Cobb
Vanderbilt University

The previous chapter detailed a considerable number of linear measurement studies and described three main themes that cut across the literature: analyzing learning in social context, accounting for the role of tools in students' development, and developing instructional design theories that examine the means of supporting learning. Stephan argued that new investigations might build on these three themes to support students' development of linear measurement conceptions. To this end, we designed and implemented an instructional sequence concerning linear measurement. Whereas the monograph itself focuses on all three themes described in chapter 2, this particular chapter focuses on the third theme, instructional design, while integrating the other two. More specifically, we expound on an instructional design theory that integrates analyses of students' learning with the design of instructional activities.

Both Cobb (chapter 1 of this monograph) and Stephan (chapter 2 of this monograph) described Design Research as an iterative process of integrating socially situated analyses of students' learning within the design of classroom environments, of which instructional sequences are one part. This approach differs from previous measurement studies in that analyses of students' learning are described both in terms of the evolving taken-as-shared mathematical development and the role of the instructional activities and tools in supporting that development. This instructional design theory served as the basis for the constitution of the measurement sequence implemented in the first-grade classroom that is the subject of this monograph. Whereas Design Research incorporates both analyses of students' learning and designing-refining instruction, this chapter focuses only on the former characteristic, building on the discussion from chapter 2 on the theoretical underpinnings of social constructivism by detailing the interpretive framework that guided the forthcoming analyses of students' measuring development (see chapter 5). We follow this discussion by describing the method used to analyze the evolving taken-as-shared learning that occurred over 32 class periods. We conclude this chapter with a description of the variety of sources used to collect data during classroom experimentation. Although we focus only on one aspect of Design Research, classroom analyses, we save chapter 4 for elaboration of the second aspect, instructional design.

ANALYSES OF STUDENTS' LEARNING

As Cobb (chapter 1 of this monograph) noted, the purpose of a design experiment is not to implement an instructional sequence and see whether it worked; rather, the purpose is to use ongoing and retrospective analyses of classroom events as fodder for improvements to the original design. The interpretive framework that guides both ongoing and retrospective analyses of students' individual and collective learning is examined in this chapter. Such a framework is necessary for understanding the actual learning that emerges as students engage in the instructional activities. The analyses, as guided by the framework, then provide feedback to inform the designer as to how the original instructional sequence supported and constrained that learning so that modifications can be made.

The theory we draw on to make sense of students' learning while we are in a classroom or when we are conducting analyses at the completion of an experiment is a version of social constructivism called the *emergent perspective*. Cobb and Yackel (1996) and Stephan (chapter 2 of this monograph) describe this theory more specifically. Briefly, this theory draws from (a) constructivist theories, which specify learning as an organic, autoregulated series of cognitive reorganizations (Steffe, von Glasersfeld, Richards, & Cobb, 1983; von Glasersfeld, 1995), and (b) interactionist theories, which emphasize learning as a social accomplishment (Bauersfeld, 1992; Blumer, 1969). These two perspectives on how learning takes place have been widely debated over the years. The emergent perspective is one attempt to transcend the individual versus social dichotomy by taking learning to be both simultaneously. In other words, learning is characterized as both an individual and a social process, with neither taking primacy over the other. Students are viewed as reorganizing their learning as they both participate in, and contribute to, the social and mathematical context of which they are a part. Motivated by this theoretical perspective, Cobb and Yackel constructed an interpretive framework useful for detailing the learning of a classroom community and its participants (see also Yackel, 1995). We subsequently elaborate how this interpretive framework guides the analyses we conduct both on a daily basis during an experiment and retrospectively.

An Interpretive Framework

Cobb and Yackel (1996) developed the framework that shapes our analyses of students' learning. This interpretive framework emerged from an attempt to conduct analyses that coordinate individual students' mathematical development with the social context of the classroom (see Figure 3.1). The left side of the framework draws on an interactionist view of communal or collective classroom processes

(Bauersfeld, Krummheuer, & Voigt, 1988; Blumer, 1969). The individual perspective draws on psychological constructivist views of students' activity as they participate in the development of these communal processes (von Glasersfeld, 1995). The relationship between the two sides of the framework is said to be reflexive.

Social Perspective	Individual Perspective
Classroom social norms	Beliefs about own role, others' roles, and the general nature of mathematical activity in school
Sociomathematical norms	Mathematical beliefs and values
Classroom mathematical practices	Mathematical conceptions

Figure 3.1. An interpretive framework for analyzing classrooms.

The social perspective consists of three aspects: classroom social norms, sociomathematical norms, and classroom mathematical practices. In an effort to explain this framework, we elaborate each of these components by focusing on the reflexivity between the social and individual perspectives.

Social Norms. The first aspect of the interpretive framework involves describing the classroom-participation structure that was established jointly by the teacher and the students. In the measurement experiment, the social norms that were interactively constituted included explaining and justifying solution methods, attempting to make sense of other students' solution methods, and asking clarifying questions whenever a conflict in interpretations arose. The teacher and the students we worked with had already negotiated these social norms before the classroom teaching experiment began, because we had worked with this classroom teacher on prior experiments, and the teacher thus saw value in initiating the negotiation of these particular social norms in her classroom. We began the classroom teacher experiment in February, four to five months into the school year, so these norms had already been relatively well established. As the teaching experiment progressed and new situations arose, however, these social norms were renegotiated. For example, two days after the beginning of the measurement sequence, the students themselves initiated a discussion about asking clarifying questions. In the episode that follows, a pair of students had just measured a portion of the classroom floor with a strip of paper containing a traced record of five of their feet—a footstrip—by placing it down end to end four times. In the middle of a lengthy whole-class discussion, Melanie explained to the class that 20 feet is the same length as four strips:

T*: Ohhh, so 20 is the number of feet, but it's only four footstrips, it's not 20 footstrips . . . it's just 4. How many people agree with what Melanie says? [Some people raise their hands.] How many people have a question for her? If you disagree? [No hands are raised.] Alice?

Alice: Well, I have a question for the whole class. Umm, but not for the people who, umm, raised their hand. Well, if they didn't understand Melanie and what she said and agreed with what she said, then they would all have questions. [Perry raises his hand.]

Perry: It seems that Ms. Smith said, or you said, raise your hand if you agree with Melanie, what Melanie said. Some people raised their hands and some people didn't. And the, umm, somebody said raise your hand if you don't agree, and they don't answer? [He seems to mean that those people who did not understand did not indicate so.]

T: Yeah, I was just wondering about that myself. That's why I was really wanting to see what people do, because I think it's, I'm really interested and I think Ms. Smith is really interested to know how you're thinking about this. So if you're not sure about what Melanie said, or you disagree with her, you need to say so.

(**T refers throughout to the person taking responsibility for leading the discussion. At any time, this individual could have been the classroom teacher or a member of the project team.)

In this episode, Alice noticed that students who did not understand Melanie's explanation did not raise their hands to ask a clarifying question. For her, these students were violating a social norm, and she explicitly challenged them about their obligations. Perry, for his part, reiterated Alice's concern when he asked why those people who "don't agree . . . don't answer" (i.e., they did not raise their hands). The teacher capitalized on this opportunity by repeating that it was important for students to vocalize their lack of understanding. As the classroom teaching experiment progressed, the social norm of asking clarifying questions became taken-as-shared.

The individual correlate to social norms concerns the teacher and students' individual beliefs about their own and others' roles. At the beginning of the teaching experiment, most students appeared to regard their role as explicating their solution methods, attempting to make sense of the solution methods of others, and asking clarifying questions when they did not understand. As illustrated by the foregoing episode, several students initiated conversations when they believed a social norm had been violated.

Sociomathematical Norms. The second aspect of the interpretive framework is describing the sociomathematical norms that were interactively constituted over the course of the teaching experiment. Cobb and Yackel (1996) discussed several sociomathematical norms including what counts as a different, an efficient, a sophisticated, and an acceptable mathematical solution. The most significant instance in which a sociomathematical norm was renegotiated during the measurement teaching experiment concerned the norm of what counts as an acceptable mathematical solution. Initially, the students explained their solution methods by describing the calculational methods that led them to a result or answer. In Thompson, Philipp, Thompson, and Boyd's (1994) terms, students were giving explanations that were consistent with a calculational orientation. Thompson and his colleagues argue that solutions reflecting a calculational orientation focus mainly on the explanation of a procedure. An alternative orientation would be conceptual in nature if a student explained his or her interpretation of the problem situation that led to a particular calculation. Thompson and his coworkers emphasize that explanations involving a conceptual orientation are more productive for mathematical discussions and for learning. As a consequence of that research, the teacher attempted to initiate a renegotiation of the criteria by which an explanation was judged as acceptable. The teacher wanted students' explanations to focus on their interpretations of task situations as well as on calculational methods (see Cobb, chapter 1 of this monograph, for a similar discussion). For example, prior to the measurement experiment, the teacher tried to initiate a conceptual discussion by asking students to describe the ways in which they anticipated solving problems. In the following episode, the teacher had posed the following problem: "Lena has 11 hearts. Dick has 2 hearts. How many more hearts does Lena have than Dick?"

T: What was it we wanted to know about the problem?
Sandra: What the answer was.
T: What was it we wanted to know about the hearts?
Sandra: The answer.
T: I think we can all agree that we want to know the answer because that is what you get when you ask a question, but what kind of answer does the question want you to tell us?
Meagan: How many more hearts does Lena have than Dick.
Mitch: We are trying to find out how many more does Dick have.
T: Do other people understand what Mitch was saying? Can someone ask him a question?
Meagan: I don't understand, but I don't know what to ask. I think he needs to explain it again.

As can be seen in this excerpt, initially the students had a calculational orientation (Thompson et al., 1994) to solving problems. The purpose of their explanations was to explain their method for obtaining "the answer." The students were not familiar with the type of explanations they were now expected to give. In fact, students like Meagan did not know how to ask questions to make sense out of such explanations. Thus, the teacher attempted to *explicitly* renegotiate what counted as an acceptable explanation just prior to the measurement sequence. Two weeks after the foregoing example, explanations that focused on conceptual interpretations of the task that led to the calculational method rather than solely the method itself became taken-as-shared.

In the following excerpt, Porter had just explained his solution method to the following problem: "Tom had 15 sandwiches. Len had 6 sandwiches. How many more did Tom have than Len?" The following excerpt illustrates a shift in the types of explanations that were being given:

T: How did you think about what happened with the sandwiches? Hilary?

Hilary: I thought about drawing a picture. I was thinking about Tom's 15 sandwiches. [She draws 15 squares to stand for sandwiches.] I got 6 out of it, and then I circled 6 and then I counted how many more were not in the circle with the 6 and I knew how many. Eight. [*She accidentally circles 7 instead of 6*].

Melanie: The ones that she circled is 7.

Hilary: Thank you, now I got it better. It would equal 9.

This excerpt illustrates that explanations were now cast in terms of students' interpretation of the problem situation, which led them to a calculation, rather than in terms of their focusing strictly on the calculational method. In fact, many students now drew pictures to describe how they reasoned about the problem. Drawing pictures not only helped students explain their reasoning but also served as a means for other students to understand their explanation. Thus, what constituted an acceptable explanation involved describing both the calculational method and the underlying interpretation of the situation.

The individual correlate of sociomathematical norms involves students' beliefs about what types of explanations are acceptable, efficient, sophisticated, and different. These beliefs are reflexively related to sociomathematical norms in that students play an active role in negotiating the criteria by which solutions are judged as acceptable. That is, as the teaching experiment progressed and new instructional situations arose, the teacher and students continually modified the criteria by which solutions were judged as acceptable.

Classroom Mathematical Practices. The final aspect of the interpretive framework involves documenting both the collective mathematical development of a classroom community and the learning of individual students. Such an analysis involves detailing the classroom mathematical practices that evolved over the course of the teaching experiment while documenting individual students' mathematical growth as they participated in, and contributed to, the emergence of these mathematical practices. A mathematical practice can be described as a taken-as-shared way of reasoning and arguing mathematically (Cobb, Stephan, McClain, & Gravemeijer, 2001). Classroom mathematical practices evolve as the teacher and students discuss situations, problems, and solution methods and often include aspects of symbolizing and notating (Cobb, Gravemeijer, Yackel, McClain, & Whitenack, 1997). To be clear, the term *mathematical practices* has been used widely in the literature to mean different things. Some researchers use the term to mean those practices from the mathematical community that are already established. The teachers' goal in this example is to help students appropriate a certain set of mathematical practices. In this monograph, we do not use the term in this manner. As we use the term, classroom mathematical practices are more localized to the classroom and are established jointly by the students and the teacher through discussion; they emerge from the classroom rather than come in from the outside. We acknowledge that in most situations, the local classroom mathematical practices are established as students bring in their ways of participating in outside mathematical practices. Hence, an interaction between outside and local mathematical practices is possible (cf. Whitenack, Knipping, and Novinger, 2001). Classroom mathematical practices differ from social and sociomathematical norms in that the latter are constructs that describe the participation structure of the classroom; that is, they describe what is normative in terms of how the teacher and the students will communicate with one another. Classroom mathematical practices, therefore, can better be thought of as the mathematical interpretations that become normative through these interactions—practices that are content specific (e.g., normative *measuring* interpretations and methods) and do not refer to more general social practices.

Individual students' mathematical interpretations and actions constitute the individual correlates of the classroom mathematical practices. Their interpretations and the mathematical practices are reflexively related in that students' mathematical development occurs as they contribute to the constitution of the mathematical practices. Conversely, the evolution of mathematical practices does not occur apart from students' reorganization of their individual activity (Cobb et. al., 1997). Here the prior cognitive analyses of Piaget become important to us. Although we have argued that prior measurement studies were primarily cast as individualistic and have advocated for analyses that discuss the role of social context in learning, we find particular value in Piaget, Inhelder, and Szeminska's (1960) cognitive constructs for analyzing students' interpretations in the case studies that appear in chapter 5 of this monograph. Specifically, we found Piaget and others' definition of measurement as the mental activity of partitioning and iterating to be more helpful for analysis than analyzing students' behaviors during measuring, in other words, learning whether a student can measure with a ruler correctly. Furthermore, we incorporated Piaget and others' mention that full measurement understanding means that students form inclusion relations among subsequent iterations, and we renamed this idea *accumulation of distance* (cf. Thompson & Thompson, 1996). In this way, we built on the cognitive analyses in the literature review to strengthen our analysis of students' development as it occurred in social context.

The primary focus of chapter 5 of this monograph is to present a sample analysis to illustrate the evolution of classroom mathematical practices, because this is the least developed aspect of the interpretive framework. Our unit of analysis will therefore be that of a classroom mathematical practice and students' diverse ways of participating in, and contributing to, its constitution (Cobb and Bowers, 1999; Cobb et al., 2001). In making reference to both communal practices and individual students' reasoning, this analytic unit captures the reflexive relation between the social and individual perspectives on mathematical activity. As difficult as it is to document the evolution of students' learning from an individualistic perspective, an even more daunting task is to chronicle the collective learning. Part of the difficulty lies in the fact that countless, complex interactions occur among a large number of students. A logical question is, How does one justify the claim that a particular mathematical interpretation is taken-as-shared in a classroom community? In the next section, we elaborate our method for analyzing all data in general, and, more specifically, we establish the criteria by which we documented the emergence of taken-as-shared, classroom mathematical practices.

METHOD OF ANALYSIS

The approach we take follows Glaser and Strauss' (1967) constant comparison method as adapted to the needs of Design Research (Cobb & Whitenack, 1996). This method, like others in the interpretivist tradition, treats data as text and aims to develop a coherent, trustworthy account of their possible meanings. The method can be organized into two distinct phases, initial conjectures and mathematical practices.

Phase 1, Initial Conjectures

With regard to the actual process of analyzing data, the first phase of the method involves working through the data chronologically episode-by-episode. In doing so, we develop conjectures about the ways of reasoning and communicating that might be normative at a particular time and about the nature of selected individual students' mathematical reasoning. These conjectures, together with the evidence that supports them, are explicitly documented as part of a permanent log of the analysis. This log also records the process of testing and revising conjectures as subsequent episodes are analyzed. A useful approach has been to focus on normative argumentation to develop these conjectures about the evolution of the collective mathematical practices (e.g., Bowers, Cobb, & McClain, 1999; Cobb, 1999; Cobb, in press; Gravemeijer, Cobb, Bowers, & Whitenack, 2000; Stephan & Rasmussen, in press). We have gained insight from Yackel (1997) who, following Krummheuer (1995), used Toulmin's (1969) model of argumentation to analyze the evolution of mathematical practices (see Figure 3.2). We have simplified Toulmin's model here to draw out the aspects most relevant to documenting mathematical practices. For a more complete version that includes rebuttals and qualifiers, see van Eemeren et al. (1996).

Figure 3.2 illustrates that for Toulmin (1969), an argument consists of at least three parts, called the core of an argument (see dotted oval in Figure 3.2): the data, the claim, and the warrant. In any argument, the speaker makes a *claim* and, if challenged, can present evidence, or *data*, to support that claim. The data typically consist of facts that lead to the conclusion. Even so, a listener may not understand what the particular data presented have to do with the conclusion that was drawn. In fact, the listener may challenge the presenter to clarify why the data lead to the conclusion. When this type of challenge is made and the presenter clarifies the role of the data in making the claim, the presenter is providing a *warrant*. Toulmin argues that a warrant is always present in an argument, whether explicitly articulated—

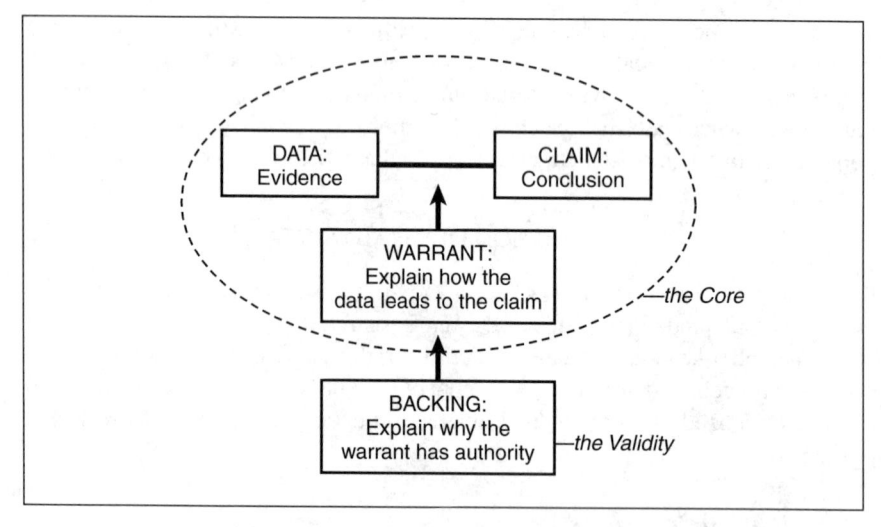

```
┌─────────────────────────────────────────────────┐
│         ┌──────────┐      ┌──────────┐          │
│         │ DATA:    │──────│ CLAIM:   │          │
│         │ Evidence │      │Conclusion│          │
│         └──────────┘      └──────────┘          │
│                   ↑                             │
│           ┌──────────────────┐                  │
│           │ WARRANT:         │                  │
│           │ Explain how the  │── the Core       │
│           │data leads to claim│                 │
│           └──────────────────┘                  │
│                   ↑                             │
│           ┌──────────────────┐                  │
│           │ BACKING:         │                  │
│           │ Explain why the  │── the Validity   │
│           │warrant has authority│                │
│           └──────────────────┘                  │
└─────────────────────────────────────────────────┘
```

Figure 3.2. A portion of Toulmin's model of argumentation.

usually as the result of a challenge—or implied by the presenter. As Toulmin's model shows, another type of challenge also can be made to an argument. The listener may understand why the data support the conclusion but may not agree with the content of the warrant used. In other words, the authority of the warrant can be challenged, and the presenter must provide a *backing* to justify why the warrant and, therefore, the core of the argument is valid.

Yackel (1997) and Cobb et al. (2001) have shown the usefulness of Toulmin's model for analyzing collective mathematical learning by relating it to the documentation of mathematical practices. In the case of the sample measurement analysis that we present in chapter 5, the *data* in students' arguments may consist of the manner in which a student measures the physical extension of an object, and the *claim* is the numerical measure of that physical extension. An appropriate *warrant* that serves to explain why the data support the conclusion might involve demonstrating how and why a measurement tool was used to produce the numerical result. This warrant can, of course, be questioned, thus obliging the student to give a backing that indicates why the warrant or measurement procedure should be accepted as having authority. A *backing* that might be treated as legitimate could involve explaining how the use of the measurement tool structures the physical extension of the object into quantities of length.

As part of the social norms that the teacher and class established, students in the first-grade classroom were expected to challenge one another's solutions when they did not understand. Often, challenges arose when new measuring tools were used. When challenges arose, students were obliged, with the teacher's help, to provide backings for why their mathematical interpretation should be accepted as valid. This need for a backing indicates that the way in which the use of a new tool structured the physical extent of objects into quantities was not taken-as-shared. In subsequent episodes, however, giving backings to justify their mathematical interpretation of a new measuring tool or measuring interpretation was no longer necessary for students. Shifts of this type are one indication that a taken-as-shared meaning for the use of the tool has become established. In other words, when a backing is no longer necessary to justify a particular measuring interpretation or a student calls for a backing and the class questions the need for one, we claim that a mathematical idea has become taken-as-shared. We therefore search for instances in the public discourse in which students no longer provide backings for a new measurement interpretation, and we record these occasions in the log for Phase 1.

We test our conjectures, for example, that a backing has dropped out and a particular measurement idea is taken-as-shared, as we analyze subsequent videotaped episodes by looking for occasions in which a student's activity appears to be at odds with a proposed normative meaning and examine how members of the classroom community treat the student's contribution. If the community does not accept the contribution, on the one hand, then we have further evidence that an idea has been established. If, on the other hand, the contribution is treated as legitimate, we need to revise our conjecture.

Phase 2, Mathematical Practices

The result of the first phase of the analysis is a chain of conjectures, refutations, and revisions that is grounded in the details of the specific episodes. In the second phase of the analysis, the log(s) of the first phase becomes data that are meta-analyzed to develop succinct yet empirically grounded chronologies of the mathematical learning of the classroom community and of selected individual students. During this phase of the analysis, the conjectures developed during the first phase about the possible emergence of taken-as-shared mathematical ideas are scrutinized from a relatively global viewpoint that looks across the entire teaching experiment. The results of the analysis are then organized and cast in terms of the analytic units that were alluded to previously—mathematical practices—and students' diverse ways of participating in them.

In summary, documenting classroom mathematical practices involves working through contributions to classroom discourse individually and using Toulmin's model of argumentation to develop conjectures about the mathematical ideas that emerge and eventually become taken-as-shared during various discussions. Once these numerous ideas are recorded, they are tested against one another and organized around common mathematical activities to develop the classroom mathematical practices. In this sense, we say that the mathematical practices are established by the community and not dropped in from outside, that is, predetermined by the designer, fully formed. Rather, our documentation of mathematical practices involves analyzing student-teacher interaction as it occurs and as the class decides what becomes legitimate mathematics in their classroom. (Stephan and Rasmussen [in press] recently extended this analysis technique by analyzing the mathematical practices that emerged in a class studying differential equations.)

To analyze the communal practices and situate students' learning in those practices, we draw from a vast amount of data. The data corpus for an analysis of this type consists of videotape recordings of all classroom sessions conducted during a teaching experiment, videotape recordings of student interviews conducted before and after the teaching experiment, and audiotape recordings of every project meeting held prior to, and during, the teaching experiment. Each of these data sources is described subsequently.

DATA CORPUS

The first-grade classroom that was the subject of this study was one of four first-grade classrooms at a private school in Nashville, Tennessee. The class consisted of 16 children: 7 girls and 9 boys. The majority of the students were from middle-class backgrounds. The classroom teaching experiment took place over a 4-month period from February to May 1996. Just prior to the teaching experiment, the teacher and the students had been engaged in instruction on single-digit addition and subtraction. The teacher was an active member of the research team and continually worked at developing a teaching practice consistent with the reform guidelines of *Professional Standards for Teaching Mathematics* (NCTM, 1991). She had been involved with the project members for 3 years and originally sought help from professionals in the mathematics education field because she had become dissatisfied with the current mathematics textbooks. The research team had developed a close professional relationship with this teacher, such that the researchers occasionally inserted commentary and questions during whole-class discussions. The students accepted the researchers' contributions as if they were visiting teachers. In the sample episodes already presented in this paper and throughout the remainder of the monograph, we use the word *teacher* collectively to refer to both the teacher and the researcher. The classroom teacher, however, took primary responsibility for leading all instruction. She clearly shared the research goals and the importance of basing instructional decisions on students' understanding. As such, we consider her an irreplaceable member of the research team.

Each of the data sources elaborated subsequently contributed in its own way to the analyses that will be presented in the next two chapters of the monograph. Data were collected from three distinct sources: classroom sessions, formal and informal meetings with the teacher, and two sets of interviews focusing on both measurement and thinking strategies for single- and two-digit addition and subtraction. Data collected from each of these sources were used to document individual students' learning (see case studies in chapter 5 of this monograph), the communal learning (see mathematical practices in chapter 5 of this monograph), and the emerging tool use (see chapter 6 of this monograph).

Classroom Sessions

Data from classroom sessions were collected in five modes: videotape recordings, field notes, clarifying questions, reflective journals, and student work. All mathematics lessons were recorded on videotape daily using two cameras. One camera was placed at the back of the classroom and recorded the teacher and all teacher-student activity taking place at the white board at the front of the room. The second camera was placed at the front of the room and recorded the students' activity during whole-class discussions.

Two researchers documented a total of five target students' individual development by asking their target students clarifying questions during pair or individual work on a daily basis. Each of these interactions with target students was videotaped to infer shifts in their individual mathematical development. Further, at the end of each day, both researchers reflected on their target students' activities and, in the form of reflective journals, made informal conjectures concerning their mathematical activity. All 16 of the students' written work was collected daily. The intent here was to infer shifts in notational and thinking strategies.

Researchers collected three sets of field notes to document instructional activities and to record whole-class discussions. On completion of the classroom teaching experiment, copies of each set of field notes were made and were discussed for purposes of triangulation.

Formal and Informal Meetings

After each classroom session, the research team met with the classroom teacher for approximately 10 to 15 minutes to plan instructional activities for the following day. Often during these informal meetings, the team offered observations of students' reasoning and made conjectures about individual and collective mathematical activity. In addition, more formal weekly meetings were held in which the pedagogical goals were revisited and the students' mathematical activity was discussed in terms of the conjectured learning trajectory. This type of discussion included how the measurement sequence was being realized in interaction and ways of further supporting the students' development. Consequently, instructional activities were revised in light of the students' activity during that week. These weekly meetings typically began by team members' sharing observations of the students' activity and making conjectures about the communal mathematical development. Both formal and informal meetings were audiotaped to document the intent of the instructional sequence as it evolved during the course of the teaching experiment and to record conversations in which we conjectured about students' development.

Formal Interviews

All 16 students were individually interviewed using two types of interview tasks. At the outset of the teaching experiment, our focus was on supporting students' development of increasingly sophisticated strategies for mental estimation and computation. Therefore, one set of interviews focused on students' thinking strategies for single-digit and two-digit addition and subtraction. As noted previously, however, our focus became that of supporting students' increasingly sophisticated measuring conceptions. Therefore, a second set of interviews centered on students' measuring conceptions. The arithmetic tasks were posed in both pretest and postinterview situations, whereas the measurement interviews were conducted as midinterviews and postinterviews. As soon as we learned how conceptually demanding learning to measure is, we developed the measurement interviews and conducted them 3 weeks after instruction on measurement had begun and immediately following the experiment.

The purpose of the pre-interviews concerning mental estimation and computation was to assess students' mathematical understanding so that the instructional sequence could build on their current ways of knowing. A second set of interviews was conducted on completion of the teaching experiment and again focused on children's thinking strategies. The intent of this postinterview was to assess students' problem-solving strategies to document growth over the 4-month period. The tasks from pre-interviews and postinterviews included solving horizontal number relations, organizing unstructured quantities, organizing structured quantities, counting by 1's, estimating, and solving story problems.

The measurement interviews were conducted, as mentioned, approximately 3 weeks after the measurement sequence began. At that time, the five target students and three other students were interviewed. These interviews were conducted to assess initial interpretations of their measuring activity. After the completion of the teaching experiment, we conducted a second set of postinterviews that focused on all 16 students' conceptions of measurement. The interview tasks focused on students' measuring activity with tools used during the teaching experiment as well as novel measurement tools.

For all sets of interviews, at least two researchers were present. One researcher conducted the interview by selecting problems from an interview protocol sheet. A second researcher recorded the interview with videotaping equipment and recorded the students' solution procedures on a protocol sheet.

CONCLUSION

In this chapter, we have detailed the interpretive framework that characterizes a student's learning as both an individual and a social endeavor. The framework enables researchers to account for individual learning by viewing it as an act of social participation. The goal of the analysis is to focus on how various students participate in, and consequently contribute to, the establishment of communal mathematical practices. To elaborate on how we identified these mathematical practices, we drew on Toulmin's (1969) model of argumentation to identify shifts in the mathematical ideas that became taken-as-shared over the course of the teaching experiment. Finally, we elaborated each of the sources of data from which the subsequent analyses were documented. In the next chapter, we focus on the second aspect of Design Research, designing testable classroom learning environments.

REFERENCES

Bauersfeld, H. (1992). Classroom cultures from a social constructivist perspective. *Educational Studies in Mathematics, 23*, 467–481.

Bauersfeld, H., Krummheuer, G., & Voigt, J. (1988). Interactional theory of learning and teaching mathematics and related microethnographical studies. In H-G. Steiner & A. Vermandel (Eds.), *Foundations and methodology of the discipline of mathematics education* (pp. 174–188). Antwerp, Belgium: Proceedings of the TME Conference.

Blumer, H. (1969). *Symbolic interactionism: Perspectives and method.* Englewood Cliffs, NJ: Prentice-Hall.

Bowers, J., Cobb, P, & McClain, K. (1999). The evolution of mathematical practices: A case study. *Cognition and Instruction, 17*(1), 25–64.

Cobb, P. (in press). Conducting classroom teaching experiments in collaboration with teachers. In R. Lesh & E. Kelly (Eds.), *New methodologies in mathematics and science education.* (pp. 307–333). Dordrecht, Netherlands: Kluwer.

Cobb, P. (1999). Individual and collective mathematical development: The case of statistical data analysis. *Mathematical Thinking and Learning, 1*, 5–44.

Cobb, P., & Bowers, J. (1999). Cognitive and situated perspectives in theory and practice. *Educational Researcher, 28*(2), 4–15.

Cobb, P., Gravemeijer, K., Yackel, E., McClain, K., & Whitenack, J. (1997). Symbolizing and mathematizing: The emergence of chains of signification in one first-grade classroom. In D. Kirshner & J. A. Whitson (Eds.), *Situated cognition theory: Social, semiotic, and neurological perspectives* (pp. 151–233). Hillsdale, NJ: Erlbaum.

Cobb, P., Stephan, M., McClain, K., & Gravemeijer, K. (2001). Participating in mathematical practices. *Journal of the Learning Sciences, 10*(1, 2), 113–163.

Cobb, P., & Whitenack, J. (1996). A method for conducting longitudinal analyses of classroom video-recordings and transcripts. *Educational Studies in Mathematics, 30*, 213–228.

Cobb, P., & Yackel, E. (1996). Constructivist, emergent, and sociocultural perspectives in the context of developmental research. *Educational Psychologist, 31*, 175–190.

Glaser, B. G., & Strauss, A. L. (1967). *The discovery of grounded theory: Strategies for qualitative research.* New York: Aldine.

Gravemeijer, K., Cobb, P., Bowers, J., & Whitenack, J. (2000). Symbolizing, modeling, and instructional design. In P. Cobb, E. Yackel, & K. McClain (Eds.), *Symbolizing and communicating in mathematics classrooms: Perspectives on discourse, tools, and instructional design* (pp. 225–273). Mahwah, NJ: Erlbaum.

Krummheuer, G. (1995). The ethnography of argumentation. In P. Cobb & H. Bauersfeld (Eds.), *The emergence of mathematical meaning: Interaction in classroom cultures* (pp. 229–269). Hillsdale, NJ: Erlbaum.

National Council of Teachers of Mathematics (NCTM). (1991). *Professional Standards for Teaching Mathematics.* Reston, VA: NCTM.

Piaget, J., Inhelder, B., & Szeminska, A. (1960). *The child's conception of geometry.* New York: Basic Books.

Steffe, L. P., von Glasersfeld, E., Richards, J., & Cobb, P. (1983). *Children's counting types: Philosophy, theory, and application.* New York: Praeger Scientific.

Stephan, M., & Rasmussen, C. (in press). Classroom mathematical practices in differential equations. *Journal of Mathematical Behavior.*

Thompson, A. G., Philipp, R. A., Thompson, P. W., & Boyd, B. (1994). Calculational and conceptual orientations in teaching mathematics. In Douglas B. Aichele (Ed.), *Professional Development for Teachers of Mathematics,* 1994 Yearbook of the National Council of Teachers of Mathematics (NCTM) (pp. 79–92). Reston, VA: NCTM.

Thompson, A., & Thompson, P. (1996). Talking about rates conceptually, part II: Mathematical knowledge for teaching. *Journal for Research in Mathematics Education, 27*, 2–24.

Toulmin, S. (1969.) *The uses of argument*. Cambridge, England : Cambridge University Press.

van Eemeren, F., Grootendorst, R., Henkemens, F., Blair, J., Johnson, R., Krabbe, E., et al. (Eds.) (1996). *Fundamentals of argumentation: A handbook of historical backgrounds and contemporary developments* (pp. 129–160). Mahwah, NJ: Erlbaum.

von Glasersfeld, E. (1995). *Radical constructivism: A way of knowing and learning*. Bristol, PA: The Falmer Press.

Whitenack, J., Knipping, N., & Novinger, S. (2001). Coordinating theories of learning to account for second-grade children's arithmetical understandings. *Mathematical Thinking and Learning, 3*(1) 53–85.

Yackel, E. (1995). *The classroom teaching experiment*. Unpublished manuscript, Purdue University Calumet, Department of Mathematical Sciences.

Yackel, E. (1997, April). *Explanation as an interactive accomplishment: A case study of one second-grade mathematics classroom*. Paper presented at the annual meeting of the American Educational Research Association, Chicago.

Chapter 4

A Hypothetical Learning Trajectory on Measurement and Flexible Arithmetic

Koeno Gravemeijer
Freudenthal Institute
Janet Bowers
San Diego State University
Michelle Stephan
Purdue University Calumet

In chapter 1, Cobb described Design Research as an approach that integrates instructional design and classroom-based experimentation in such a manner that the design effort shapes the research effort and the research effort shapes further design as well (see also Gravemeijer, 1998). In chapter 2, Stephan examined the theoretical perspective that underlies the research effort. And in chapter 3, Stephan and Cobb explored the interpretive framework and method that serve as the basis for analyzing the ongoing events of a classroom environment. In this chapter, we first focus on the theoretical underpinnings that guided the process of instructional design for the measurement experiment. We then cast this perspective in terms of an instructional theory on measuring and flexible arithmetic that guided the present teaching experiment.

DESIGN RESEARCH

The type of Design Research in which we engage falls under the broad headings of *developmental research* (Cobb, Gravemeijer, Yackel, McClain, & Whitenack, 1997; Freudenthal, 1991; Gravemeijer, 2001; Gravemeijer, 1998) and *transformational research* (NCTM Research Advisory Committee, 1996). Other users of the term *Design Research* (Brown, 1992; Collins, 1999) define it in an interpretation similar to ours. At the heart of Design Research is a cyclic process of designing instructional sequences, testing and revising them in classroom settings, and then analyzing the learning of the class so that the cycle of design, revision, and implementation can begin again. Therefore, testing and revising occur at both a formative level—that is, on an ongoing, daily basis during classroom intervention—and a summative level—that is, retrospectively. At the core of this cycle is the interplay between designing a potential instructional sequence and using analyses of students' learning while they are engaged in the instructional activities to shape the redesign of the sequence. In what follows, we elaborate on the general heuristics that guide the development of an instructional sequence in the design phase.

Realistic Mathematics Education

The preliminary design and continual modification of an instructional sequence is guided by the domain specific instructional theory called Realistic Mathematics Education (RME), developed at the Freudenthal Institute (Gravemeijer, 1994a; Streefland, 1991; Treffers, 1987). RME is consistent with the emergent perspective in that both are based on a view of mathematics as a human activity and a view of mathematical learning as a process that occurs as students develop ways to solve problems within the social context of a mathematics classroom (Cobb, 2000). Given these assumptions, the main activity that designers attempt to support is progressive mathematization, or level-raising: Students' activity on one level is subject to reflection and analysis on the next level. Although many important ideas are embodied within the idea of mathematization, we have chosen to highlight the three basic RME heuristics that, when taken together, inform designers in their efforts to support students' reasoning within the cycle of Design Research.

Heuristic 1. Sequences must be experientially real for students. One of the central heuristics of RME is that the starting points of instructional sequences should be experientially real in that students must be able to engage in personally meaningful activity (Gravemeijer, 1994b). Often, this requirement means grounding students' initial mathematical activity in experientially real scenarios, which can include mathematical situations. To find such scenarios, the designer may carry out a didactical phenomenological analysis, within which he or she investigates what phenomena are organized by the mathematical concepts or procedures to be developed by students. For example, a scenario that was developed for the first-grade teaching experiment was one in which a king needed to measure the lengths of particular items in his kingdom. Of course, by saying the starting points are *experientially real*, we are not suggesting that the students had to experience these starting points firsthand. Instead, we are saying simply that the students should be able to imagine acting in the scenario, as king, for example.

Heuristic 2: Guided Reinvention. A second heuristic of RME involves what Freudenthal (1991) termed *guided reinvention*. To start developing a sequence of instructional activities, the designer first engages in a thought experiment to imagine a route the class might invent (Gravemeijer, 1999). Here, knowledge of the history of mathematics as well as prior research concerning students' invented mathematical strategies can be used to develop what Simon (1995) has called a Hypothetical Learning Trajectory, or a possible taken-as-shared learning route for the classroom community. This construct is further elaborated in a subsequent section of this chapter.

Heuristic 3. Emergence of student-developed models. A final heuristic of Design Research involves building on students' informal modeling activity to support the reinvention process. As such, instructional sequences should provide settings in which students can model their informal mathematical activity (Gravemeijer, 1994a). Often, therefore, the designer imagines what types of practices will emerge as students reason with certain tools—physical tools, symbols, or notation—that might support taken-as-shared development of new mathematical solutions. Integral to this heuristic is the conjecture that students' *models of* their informal mathematical activity evolve into *models for* increasingly sophisticated mathematical reasoning. Gravemeijer (1999) notes that although a model may manifest itself in a variety of ways during the sequence, the model is a global notion that spreads over the instructional sequence.

The interplay among the three heuristics: Anticipating a transition from model of *to* model for. In the RME approach, all three heuristics are seen as working in unison to support the development of increasingly sophisticated mathematical practices as students participate in the sequence of activities. In what follows, we describe this unison by first elaborating what is meant by a *model* from the perspective of an RME developer. Next, we describe the *model of* to *model for* transition that is the center of this approach.

The word *model* can be interpreted in different ways. Specifically, RME's interpretation differs from that of mainstream information-processing approaches or cultural-historical approaches, such as that elaborated by Gal'perin (1969). In the latter approaches, models come to the fore as didactic models that embody the formal mathematics to be taught. In other words, the formal abstract mathematics is seen as *concretized* to be made accessible for the students through *discovery*. This view is consistent with the sociocultural view of tools described in chapter 2. In contrast with this view, RME models are not derived from the intended mathematics. These models are seen as student-generated ways of organizing their mathematically grounded activity. Thus, the RME models are products of model*ing*; the starting point is in the experientially real situation of the problem that has to be solved, for example, the king scenario. The idea is that acting with these models will help the students *(re)invent* the more formal mathematics that constitutes the goal of instruction.

According to the theory of RME, student-generated models evolve as students collectively construct new mathematical realities. Therefore, a transition in students' activity from *model of* to *model for* indicates a shift in the ways that the students are participating in and hence constituting the mathematical practices of the classroom community. In theory, the transition in models occurs as follows. Initially, the models are to be understood as context-specific models. The models refer to concrete or paradigmatic situations that are experientially real for the students. On this level, the model should allow for informal strategies that correspond with situated solution strategies at the level of the situation that is defined in the contextual problem. From then on, the role of the model begins to change. As the classroom community gathers more experience with similar problems, a taken-as-shared model becomes more general in character. The process of acting with the model changes from one of constructing solutions situated in the context to one of using the model to communicate reasoning strategies. In this sense, the model becomes an object in and of itself within the classroom community. By then, the model becomes more important as a basis for mathematical reasoning than as a way to represent a contextual problem. As a consequence, the model can become a referential base for the formal mathematics. Or, in short, a *model of* informal mathematical activity becomes a *model for* more formal mathematical reasoning.

The shift from *model of* to *model for* can be seen as aligned with shifts in the collective mathematical practices. In brief, researchers look for these shifts in practices by noting changes in students' justifications that reflect differences between explanations of their solutions within the modeled context and explanations of mathematical relations. As an aside, a helpful approach has been to limit the use of the *model of–model for* characterization to more encompassing shifts. Otherwise, every symbolizing act could be predicated as a model shift. Usually, something is symbolized (model of), and the symbolization is used to reason with (model for). If we include all such shifts in the emergent-models heuristic, this heuristic would lose its power. We therefore want to delineate when we do or do not speak of a shift from a *model of* to a *model for* by specifying that such a shift coincides with the creation of some new mathematical reality (Gravemeijer, 1999).

In summary, the three RME heuristics can be seen as integrated into the process of supporting shifts from contextualized activity to more formal ways of reasoning. In the following section, we describe the process by which designers anticipate how this transition occurs through the development of a hypothetical learning trajectory. After that, we illustrate the process by describing the hypothetical learning trajectory that was developed for the measurement and number-relations sequences.

Distinctions Between a Hypothetical Learning Trajectory and Traditional Lesson Planning

The RME design heuristics described in the foregoing serve as tools for enabling instructional designers to conceptualize how a teacher might build on students' models to support their construction of increasingly sophisticated reasoning. One question that often arises is the inherent tension between adapting one's teaching to the ideas and suggestions of the students and, at the same time, planning instructional activities in advance. Our response to this apparent paradox of "on-the-fly planning" is to revisit Simon's (1995) study of the role of a *constructivist* teacher. Simon analyzed his own role as a teacher who tries to influence his class's argumentation by playing off their personally developed mathematical meanings. He tries to envision ways in which his students will engage in the activities, then tries to anticipate how his students' potential lines of argumentation might lead to the types of explanations and justifications he wants to become taken-as-shared in the evolving classroom community. To describe this process, Simon introduces the notion of a *hypothetical learning trajectory* (HLT):

> The consideration of the learning goal, the learning activities, and the thinking and learning in which the students might engage make up the hypothetical learning trajectory .... (Simon, 1995, p. 133)

We have come to believe that this notion of an HLT addresses the apparent paradox of "on-the-fly planning" by taking into account four specific considerations of the process that distinguish it from the traditional process of lesson planning: (1) the socially situated nature of the learning trajectory, (2) the view of planning as an iterative cycle rather than a single-shot methodology, (3) the focus on students' constructions rather than mathematical content, and (4) the possibility of offering the teacher a grounded theory describing how a certain set of instructional activities might play out in a given social setting. Each of these considerations is further elaborated subsequently.

The first difference between the RME approach and that of traditional lesson planning is that RME designers view learning as situated within a classroom activity system. Two implications of this assumption are that we do not assume that any given sequence will play out the same in any classroom, and we do not view a proposed trajectory as a series of conceptual stages along which each individual student in the class will progress. Instead, when developing a learning trajectory, we attempt to outline conjectures about the collective development of the mathematical community by focusing on the practices that might emerge at the beginning of the sequence, then creating tools and activities that might support the emergence of other practices that would be based on increasingly sophisticated ways of acting and justifying mathematical explanations.

The second aspect of the RME planning process that distinguishes it from more traditional teacher-only lesson planning is that it incorporates feedback from research and, hence, is cyclic in nature. As the teacher and students engage in the activities, new insights into the types of mathematical practices that are emerging form the basis for the constitution of a modified HLT for the subsequent lessons. Simon (1995) describes this process as a *mathematics teaching cycle*. Although the teaching cycles that Simon used as examples concerned the teacher's preparation for the next day's instructional activity, we have extended this notion of cyclic planning to characterize the process by which the teacher envisions a more encompassing learning route. Thus, when enacting any activity, the teacher adjusts her or his time to build on what actually happens in the classroom. The overall plan provides a basis for thinking through how to get back on track. In relation to this flexibility, we may borrow Simon's journey metaphor. When making a journey, we may start out with a well-considered travel plan. Nevertheless, our actual journey will differ from this plan because of the circumstances that we meet during our journey. Moreover, not every aspect of a journey can be planned in advance. A preconceived, conjectured learning route may be compared with a travel plan, in which what is actually going on in the classroom constitutes the journey.

A third characteristic of the design process that distinguishes it from traditional lesson planning and also resolves the apparent "on-the-fly planning" paradox is that the HLT takes students' cognitive development as the basis for design. This focus stands in stark contrast with traditional instructional design programs that organize the goals of instruction in terms of mathematical content. Therefore, Simon's notion of an HLT is more consistent with reform recommendations that place high priority on students' mathematical reasoning and justifications (NCTM, 2000).

The final distinction between Design Research and traditional lesson planning is that the result of a Design Research experiment, like the one that is the topic of this monograph, can be cast in terms of a grounded theory describing a learning route. The latter is also referred to as a conjectured instructional theory. The term *instructional theory* is used to convey the intent of offering a socially situated rationale. That is, the objective of Design Research is not to offer an instructional sequence that "works" but to offer a grounded theory that describes the tools, imagery, discourse, and mathematical practices that emerge as the instructional sequence is played out in the social setting of a mathematics classroom.

This last criterion, offering a grounded description of how a sequence will play out, is consistent with the message of a recent editorial published by the Research Advisory Committee (RAC) of the National Council of Teachers of Mathematics (NCTM) (1996). In its piece, the RAC observed that fewer research articles, at least those published in the *Journal for Research in Mathematics Education*, simply report findings that show that instructional approach A works better than approach B. The rationale for this trend is that researchers have realized that simply comparing two instructional approaches does not have much to offer to teachers who are not successful with the approved approach. In other words, the only thing that those teachers know is that it worked elsewhere. In contrast, in the instance of a justification in terms of a grounded theory about how a new approach works, teachers have a basis for thinking through why something may not have worked in their classroom and beginning to conjecture about what adaptation might be helpful. Note that, from this perspective, the enactment of an instructional sequence can be seen as a continuation of the teaching experiment, in that teachers can contribute to the research; the knowledge gained by teachers may be used to enhance the conjectured instructional theory. Moreover, teachers become aware of the social nature of learning, in that the Design Research perspective takes advantage of the diversity of classroom settings by offering a global learning route to be tailored to the specific situations by classroom teachers. In short, when we discuss the conjectured instructional theory that is the yield of the teaching experiment on linear measurement and flexible arithmetic, the reader should not expect a textbook-type elaboration of an instructional sequence. Instead, our presentation looks more like a socially situated rationale for an instructional sequence (an example will be presented in chapter 6 of this monograph).

DESIGNING THE HYPOTHETICAL LEARNING TRAJECTORY FOR THE FIRST-GRADE SEQUENCE

We begin by situating the classroom teaching experiment on measurement and arithmetic in the context of the research that preceded it. One aspect that we need to elaborate here is that originally, learning to measure was a means to an end—developing a sequence for supporting students' constructions of computational strategies for reasoning with numbers up to 100—and not a goal in itself. Learning to measure, however, proved to be a sophisticated process that deserved attention in its own right. Therefore, the actual teaching experiment encompassed two partial sequences consisting of linear measurement and mental computation.

Prior classroom-based research

One source that influenced the design of the current activities was our prior research of the empty-number-line instructional sequence. The goal of the empty-number-line sequence was to support students' activities so that they would develop flexible computation strategies for addition and subtraction up to 100 (Whitney, 1988; Treffers, 1991). In prior research settings, we found that students reasoned with this tool to describe the solution of two-digit addition and subtraction problems by marking the numbers involved and drawing "jumps" that corresponded with their partial calculations. For example, solving 64 – 29 by subtracting 4, 10, 10, and 5, respectively, might have resulted in the inscription shown in Figure 4.1. Such a solution method, which could be seen as a shortened form of counting down, belongs to the most common informal solution methods that students use.

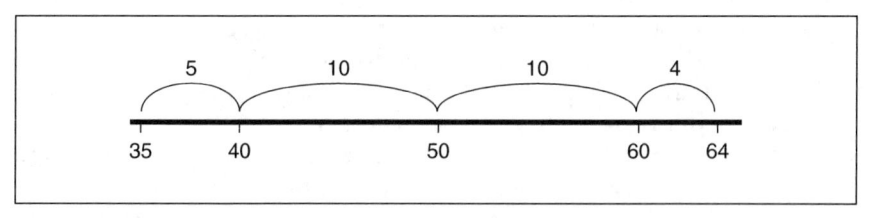

Figure 4.1. 64 – 29 on the empty number line.

In relation to this categorization, we found that we were helped in our research by distinguishing between collection-type solution methods and linear, counting-type solution methods. In the first type, students partition each number into a tens part and a ones part, then add or subtract the numbers of those parts separately. In contrast, when reasoning with linear-based methods, students increment or decrement directly from one number as in the example in Figure 4.1.

We found this distinction useful because research has shown that collection-type solutions may cause problems when used with more complex subtraction tasks (see Beishuizen, 1993). Therefore, the empty-number-line sequence was especially designed to support more counting-type reasoning strategies. We also acknowledged, however, the fact that collections-based reasoning has its merits, in relation to both the written algorithms and estimation. Therefore, we also planned to provide opportunities in which students could develop well-understood, collection-based strategies for reasoning with tens and ones. Note that our aim was not to teach the students efficient solution strategies as such. Instead, our aim was to help students develop a framework of number relations, which would enable them to construe flexible solution methods.

As part of the cycling process of Design Research, our prior research on the empty number line led us to make revisions to the arithmetic instructional sequence (Cobb et al., 1997). The original arithmetic sequence developed in the Netherlands incorporated a bead string with red and white alternating colors every ten beads as the foundational imagery for reasoning with the empty number line. The indispensability of this kind of imagery proved itself in the teaching experiment conducted by Cobb and others, in which the bead string was skipped. Without some sort of linear-based contextual imagery, alternative interpretations concerning the meaning of numerals as specified on the number line arose. For example, students interpreted "73" as either the 73rd instance of counting or as the result of having counted 73 times. In using the bead string, these interpretations are equivalent to students' deciding whether to put a clothespin, for example, directly on the 73rd space or on the space after. Without the imagery of, and discussions about, the bead string as a background, the students could not resolve these issues. Because the underlying imagery had been lacking with its introduction, the students needed to create a semantic grounding for the empty number line individually and privately. They did so in a variety of ways, rendering impossible their resolution of interpretation issues; they had no common ground. Rather than fall back on the bead string, which lacks in real-life motivation—no one would realistically place clothespins on a bead string—we designed the measurement sequence as a replacement for reasoning with a bead string. Thus, this monograph serves more broadly as one complete iteration of the Design Research Cycle: using prior analyses of the empty number line to revise and subsequently test a new instructional sequence and then analyzing and revising the enactment of the resulting instructional sequence.

As we began planning a new sequence that would precede the empty number line, we focused on the ruler as the prime model. With this model in mind, we then focused on prior research related to students' understandings of measurement. As noted in chapter 2, Piaget's prior research on children's learning revealed that to have students view measuring as an act of structuring space, they must first develop concepts of nested and iterable units. To use Greeno's (1991) terms, the intent of the instructional sequence would be that students would come to act in a spatial environment in which the results of iterating signify an accumulation of distances. Our decision to use nonconventional units was influenced by the work of Piaget, Inhelder, & Szeminska (1960); Steffe, von Glasersfeld, Richards, & Cobb (1983); and Thompson & Thompson (1996), who maintain that understanding measuring involves developing an image of accumulated distance. Whereas some researchers agree with the approach of having students build a "ruler" out of iterating nonconventional and then conventional units (Lehrer, in press), others argue that children prefer to use a conventional ruler from the beginning, and their research findings support such an approach (Clements, 1999; Nunes, Light, & Mason, 1993). Our reasoning for the choice of nonconventional units is explained in the following section.

The Hypothetical Learning Trajectory

We started our design activity with the observation that a strong similarity is evident between the empty number line and a ruler. Speculating on the genesis of the ruler in history, we reasoned that we could take the view that the ruler came about as a record of iterating a measurement unit. So the ruler could be thought of as a *model of* iterating some measurement unit. As described previously, our prior classroom-based research indicated that the empty number line could function as a *model for* mathematical reasoning in the context of mental computation strategies for reasoning with numbers up to 100. The connection between the two would be in the relation between iterating measurement units as accumulating distances and a cardinal interpretation of positions on the number line. In the context of measurement, we hoped that students would come to see distances on a ruler or number line as magnitudes. By using the same tool as a means for supporting students' efforts to solve contextual problems that do not involve measurement, we hoped students would engage in practices that involved reasoning about quantities more generally. Finally, even the direct reference to specific quantities might become less prominent, since the numbers on the empty number line would start to derive their meaning from a framework of number relations for the students.

One of the overriding concerns in this sequence was to bring the decimal structure of our number system into focus for whole-class discussions. In the context of measurement, we planned to do so by designing activities in which students would shorten their iterating activity by means of a larger measurement unit, that is, a unit of 10. Measuring with tens and ones can form the basis for the development of students' view of the ruler as structured in terms of tens and ones. At the same time, we hoped it would support the emergence of other strategies that use decuples as reference points.

In summary, we envisioned the skeleton of a sequence for supporting students' arithmetical reasoning with numbers less than 100 that would unfold as follows:

1. Constructing context-based meaning for measuring.
    a. Iterating units. The students start measuring the lengths of various objects by iterating some measurement unit.
    b. Organizing or structuring space. The students discuss their solutions in terms of structured space covered by iterated units.
2. Reasoning with a ruler. The students explain their efforts to shorten their iterations of counting by using tens and ones. While doing so, the students start to develop a framework of number relations with decuples as reference points.
3. Extending the activity of measuring to incrementing, decrementing, and comparing lengths. Here the activity shifts from measuring to reasoning about

measures. This type of task would offer students the opportunity to develop arithmetical solution methods that might be supported by referring to the decimal structure of the ruler.
4. Using a schematized ruler as a means of scaffolding and communicating arithmetical solution methods for measurement problems. The students develop increasingly sophisticated methods for arithmetical reasoning based on an emergent framework of number relations.
5. Reasoning with a flexible ruler. The students develop new ways of communicating solution methods for all sorts of addition and subtraction problems.

As Cobb (chapter 1 of this monograph) noted, the Design Researcher not only envisions a sequence of instruction but also forms conjectures about the potential mathematical argumentation and evolving tool use that might accompany the realization of the sequence. As such, the global structure of the anticipated learning trajectory would include conjectures of possible discourse and tool use. Hence, in preparation for the teaching experiment, this global structure of a hypothetical learning trajectory was worked out in the following manner.

- The sequence would start with pacing to structure linear distance. Measuring with centimeters or inches would have been so familiar to the students that the need for structuring of iterative units might not have become an explicit topic of discussion. Moreover, we imagined that measuring with one's feet would create the opportunity to raise awareness for the need for a standardized measurement unit; one such scenario introduces the king's foot as a standardized measurement unit (cf. Lubinski & Thiessen, 1996; Lehrer, 2003). We anticipated that the discourse would focus on students' ways of structuring distances as they counted paces in different manners.
- Measuring with the king's foot would be followed by measuring with a different iterable unit: Unifix cubes in a smurf scenario. According to the story that would be told to the students, the smurfs—little blue dwarfs—measure with their food cans, which happen to have the size of Unifix cubes. An important instructional design consideration here was rather practical, namely, that this smaller measurement unit would allow students to deal with larger numbers more easily.
- Next students would be confronted with a dilemma in which the smurfs decided to only use a small number of food cans for measuring instead of a bag of 50 cans. Students would offer suggestions of smaller quantities (e.g., 2, 5, 10, 20), and the teacher would choose 10 to capitalize on the decimal structure of our number system. As students would engage in measuring activities with a rod of ten cans, we anticipated that the discourse would focus on structuring distance in tens and ones and coordinating measuring with two different-sized units. Furthermore, the idea that a rod of ten would signify the result of counting Unifix cubes might become taken-as-shared. Discussions concerning the result of iterating as an accumulation of distance might emerge here in students' attempts to clarify their measuring interpretations.

- As students reasoned with these nonconventional measuring tools, we envisioned that they would begin to stop counting every unit mark and begin counting by tens and then adjusting by adding or subtracting ones. Ultimately, we hoped that measuring with tens and ones would help the students reason with tens and ones to structure the number sequence up to 100.
- Measuring by tens and ones would lay the basis for introducing a measurement strip that was made of a paper strip on which 10 units of ten were iterated, each subdivided into 10 units of one cube (Figure 4.2). The idea was that measuring with the measurement strip would become a taken-as-shared way of signifying the result of iterating with the smurf bar and the individual cubes. Discussions might center on the meaning that certain marks or numerals had for students. We could imagine accumulation of distance becoming a taken-as-shared interpretation of the symbols on the measurement strip. In other words, we hoped that eventually a number on the measurement strip would signify the distance that extends from the beginning of the measurement strip to the line to which this number belongs.

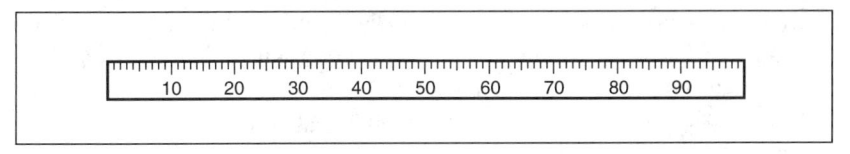

Figure 4.2. Measurement strip of 100.

- After the students grew accustomed to reasoning with the measurement strip, we envisioned that a shift might be made in taken-as-shared mathematical reasoning from measuring objects to using the measurement strip to reason about comparing, incrementing, and decrementing lengths of objects that are not physically present. The focus of discourse would be on identifying and comparing numerically the lengths and heights of objects. In this way, the discourse would shift from interpreting the result of measuring to quantifying the difference among various heights. These tasks were designed to offer opportunities for developing arithmetical solution strategies as described previously.
- To further foster a shift away from measuring to reasoning about magnitudes, another tool, the measurement stick, would be introduced along with a new scenario. In this case, the stick was attached to a wall and used to read off the height of the water in a canal (see Figure 4.3). The students would be invited to reason with the measurement stick to solve contextual problems around the rise and fall of water heights. The design of the measurement stick was based on our prior research with another tool, the bead string. An important feature of this tool was that although it has no numbers, values are easy to read off by counting tens

and ones. The conjecture was that this feature might prompt discussions in which students stop counting and instead rely on the decimal structure of the stick.

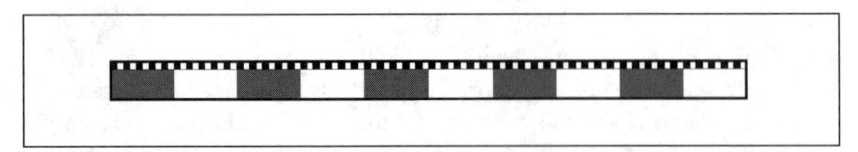

Figure 4.3. Measurement stick. (Here the stick is shown in a horizontal position; in the context of the water heights, the stick is positioned vertically.)

- Next the empty number line would be introduced. The goal of this tool was to support students' efforts to reason with a tool that was not demarcated with specific units. The idea was that these line-based solution methods would still be grounded in measuring with units of 10 and 1. We anticipated that students would draw on the practices that emerged during measuring to develop and interpret strategies with the empty number line. For example, jumps on the number line might symbolize distance quantities that themselves signify accumulation of distances. For some students, these jumps might also signify iterating Unifix cubes or rods of ten.
- As a final step, the students would be encouraged to use the same solution strategies and explanations that evolved when reasoning with the empty number line to solve contextual problems involving incrementing, decrementing, and comparing. Here we hoped that jumps on the number line would signify quantitative changes in the contextual situation under consideration. Eventually, however, we expected that the students' thinking strategies would become based on more generalized numerical relationships. The shift from numbers as referents to numbers as mathematical entities is reflexively related to the *model of–model for* transition. Initially, the students' actions with the model would foster the constitution of a framework of number relations. After the transition, in which students develop a framework of number relations, the model would become a model for generalized mathematical reasoning.

Cascade of Tools and Inscriptions

The succession of various tools constitutes an essential element of the instructional sequence. Although the ruler served as the overarching model, it was manifested in the form of various tools or inscriptions throughout the hypothetical learning trajectory (Gravemeijer, 1999). This representation is characteristic of our view that reasoning with each new tool should support the students' efforts to shorten or stop their previous activity when reasoning with earlier ones. We hoped the class would participate in a series of emergent mathematical practices in which each successive practice was based on students' increasingly sophisticated tool use and discourse. Moreover, we believed a meaningful buildup of reasoning strategies would ensure that all students had a way of participating in the practices because anyone could go back to earlier, more familiar ways of participating if the need arose. From this perspective, an essential image for the teacher to have was how the reasoning strategies that we envisioned for each tool built on previous reasoning strategies.

We find Latour's (1990) description of a "cascade of inscriptions" to be a useful way to describe how the proposed sequence of tools can be seen as reflecting RME's theoretical reinvention process. We say *theoretical* because we need to acknowledge that we cannot expect students to spontaneously invent these tools as needed at just the right time. Here, again, we rely on the teacher to resolve the apparent "on-the-fly planning" paradox. In this case, our solution was that the teacher would introduce the new tools as a way to do justice to the reinvention idea while also moving the class forward. The timing and process by which the teacher introduced each new tool were crucial. First she had to make sure that each new tool was discussed in the whole-class setting as a solution to a problem or an encapsulation of some reasoning strategy. For example, the teacher would tell a story about how the king did not want to do all the measuring by pacing each length personally. The teacher would then involve the students in this problem by asking them for their solutions (cf. Stephan, Cobb, Gravemeijer, & Estes, 2001). Next the solutions offered by the students would become the topic of a whole-class discussion in such a way that the students could explore the advantages and disadvantages of each suggested solution. Only then would the teacher inform the students of the solution that was chosen in the story. With luck, a similar solution had already been suggested by one of the students. If not, the preceding discussion would have at least have provided a basis on which the students could conclude that the teacher's proposed solution was sensible under the given conditions.

Another aspect of the teacher's role in introducing each new tool was that she needed to emphasize the tool's relation with the preceding activities. That is, the context problem for which the new tool offered a solution could not be an arbitrary problem; it had to be one that was rooted in the preceding scenario and that operated under the preceding constraints. In the instance of the king who did not want to do all the measuring, the problem had it roots in the earlier activity of measuring with the king's foot, or measuring like a king by pacing with students' feet. This aspect is important because that history must provide the imagery underlying the new tool. In the case of the unwilling smurfs, a bar of 10 cubes was introduced to build on the activity of measuring with individual cubes. More precisely, the imagery underlying the bar of 10 was that of a record of measuring with single cans. Students were expected to discuss advantages of this tool, such as the fact that it could be used as a tool for a shortened form of measuring, that is, it was easier to carry 10 cans than a whole bag, and counting by tens was simpler than counting by ones. Taken together, these three elements—introducing each new tool as a solution to a problem, collectively discussing the advantages and disadvantages of the new tool, and the importance of a context-relevant scenario—were designed to engage the students in the experience of the reinvention process.

CONCLUSION

This chapter has focused on the ways in which we were informed in our design process by the overall ideas underlying Realistic Mathematics Education. We began the chapter by describing three heuristics: experientially real activities, guided reinvention, and emergent models. We then described how each of these heuristics enabled us to address the "on-the-fly planning" paradox of how teachers can, on the one hand, adapt to the ideas and solutions of the students and, on the other hand, move the class toward a mathematical goal. Following Simon (1995), we argued that the essential component is anticipating the collective practices in which the students might engage as they reason with the various tools in the context of the activity scenarios. In the second part of the chapter, we described the hypothetical learning trajectory or conjectured instructional theory that functioned as a framework of reference as we enacted the sequence.

One of the major themes of this chapter has been the importance of considering how the sequence will support the taken-as-shared development of increasingly sophisticated thinking. The conjectured instructional theory we offered addresses this concern by first describing the general concept of modeling. We envisioned that, as students reasoned with various tools, the ruler would shift from serving as a model of iterating or measuring to being a model for arithmetical reasoning. The *model of–model for* shift is the tie that binds the measurement and arithmetic instructional sequences.

A second fundamental theme that was presented in this chapter was our view of the role of tools in supporting learning. In brief, the solutions that students develop depend heavily on the fact that they are reasoning with the tools they are given. This view differs from a view of tools as devices that enable one to transfer one's mental reasonings to a physical device. This view also differs from a view of a tool as a culturally laden device whose use and meaning are predetermined.

Teachers who want to adhere to the calls of reform mathematics education to build on students' own contributions may build on our work, but they will need to construe their own hypothetical learning trajectories on a day-to-day basis to be consistent with the anticipated practices in each social setting. Thus, teachers will need to take into consideration their knowledge of their own students, the instructional history of the class, and the end points they envision. In relation to these considerations, we can recall Simon's (1995) journey metaphor. The instructional theory offers a travel plan—or maybe a travel guide, on the basis of which one could design a travel plan—but the actual journey of the teacher and the students will always differ from the travel plan. Still, we would argue, the better the travel plan, the easier the journey.

At this point, we hope that we have piqued the reader's interest to the point of wondering how the proposed sequence actually played out. The next chapter addresses this question by presenting an analysis of how the mathematical practices actually emerged over the course of the enacted sequence and how this emergence compares with the hypothetical learning trajectory described here. In this chapter, we have presented the results of prior research and described how the analyses of that research shaped the design of the measurement-arithmetic sequence. Consistent with the Design Research Cycle, the classroom analysis found in the next chapter is used to generate possible mathematical practices that accompany the proposed instructional sequence and to provide the rationale for revisions to it. In the chapter following the analysis, we discuss revisions and outline a generalized instructional theory on measurement and linear-based arithmetic, thus completing one iteration of the Design Research Cycle.

REFERENCES

Beishuizen, M. (1993). Mental Strategies and materials or models for addition and subtraction up to 100 in Dutch second grades. *Journal for Research in Mathematics Education, 24*(4), 294–323.

Brown, A. L. (1992). Design experiments: Theoretical and methodological challenges in creating complex interventions in classroom settings. *Journal of the Learning Sciences, 2*, 141–178.

Clements, D. (1999). Teaching length measurement: Research challenges. *School Science and Mathematics, 99*(1), 5–11.

Cobb, P. (2000). Conducting classroom teaching experiments in collaboration with teachers. In R. Lesh & E. Kelly (Eds.), *New methodologies in mathematics and science education* (pp. 307–333). Dordrecht, Netherlands: Kluwer.

Cobb, P., Gravemeijer, K., Yackel, E., McClain, K., & Whitenack, J. (1997). Symbolizing and mathematizing: The emergence of chains of signification in one first-grade classroom. In D. Kirshner & J. A. Whitson (Eds.), *Situated cognition theory: Social, semiotic, and neurological perspectives* (pp. 151–233). Hillsdale, NJ: Erlbaum.

Collins, A. (1999). The changing infrastructure of educational research. In E. C. Langemann & L. S. Shulman (Eds.), *Issues in education research* (pp. 289–298). San Francisco: Jossey Bass.

Freudenthal, H. (1991). *Revisiting mathematics education*. Dordrecht, Netherlands: Kluwer.

Gal'perin, P. Y. (1969). Stages in the development of mental acts. *Tijdschrift voor nascholing en onderzoek van het reken-wiskundeonderwijs, 7*(3, 4), 22–24.

Gravemeijer, K. (1994a). *Developing realistic mathematics education*. Utrecht, Netherlands: CD-β Press.

Gravemeijer, K. (1994b). Educational development and developmental research. *Journal for Research in Mathematics Education, 25*(5), 443–471.

Gravemeijer, K. (1998). Developmental research as a research method. In J. Kilpatrick & A. Sierpinska (Eds.), *Mathematics education as a research domain: A search for identity* (an ICMI study) (book 2, pp. 277–295). Dordrecht, Netherlands: Kluwer.

Gravemeijer, K. (1999). How emergent models may foster the constitution of formal mathematics. *Mathematical Thinking and Learning, 1*(2), 155–177.

Gravemeijer, K. (2001). Developmental research: Fostering a dialectic relation between theory and practice. In J. Anghileri (Ed.), *Principles and practice in arithmetic teaching* (pp. 147–161). London: Open University Press.

Greeno, J. (1991). Number sense as situated knowing in a conceptual domain. *Journal for Research in Mathematics Education, 22*, 170–218.

Latour, B. (1990). Drawing things together. In M. Lynch & S. Woolgar (Eds.), *Representations in scientific practice* (pp. 19–68). Cambridge, MA: The MIT press.

Lehrer, R. (2003). Developing understanding of measurement. In J. Kilpatrick, W. G. Martin, & D. Schifter (Eds.), *A research companion to "Principles and standards for school mathematics"* (pp. 179–192). Reston, VA: National Council of Teachers of Mathematics.

Lubinski, C., & Thiessen, D. (1996). Exploring measurement through literature. *Teaching Children Mathematics, 2*(5), 260–263.

NCTM Research Advisory Committee (1996). Justification and reform. *Journal for Research in Mathematics Education, 27*(5), 516–520.

National Council of Teachers of Mathematics (NCTM) (2000). *Principles and standards for school mathematics.* Reston, VA: NCTM.

Nunes, T., Light, P., & Mason, J. (1993). Tools for thought: The measurement of length and area. *Learning and Instruction, 3*, 39–54.

Piaget, J., Inhelder, B., & Szeminska, A. (1960). *The child's conception of geometry.* New York: Basic Books.

Simon, M. A. (1995). Reconstructing mathematics pedagogy from a constructivist perspective. *Journal for Research in Mathematics Education, 26*, 114–145.

Steffe, L. P., von Glasersfeld, E., Richards, J., & Cobb, P. (1983). *Children's counting types: Philosophy, theory, and application.* New York: Praeger Scientific.

Stephan, M., Cobb, P., Gravemeijer, K., & Estes, B. (2001). The roles of tools in supporting students' development of measuring conceptions. In A. Cuoco (Ed.), *The roles of representation in school mathematics* (2001 Yearbook) (pp. 63–76). Reston, VA: National Council of Teachers of Mathematics.

Streefland, L. (1991). *Fractions in realistic mathematics education. A paradigm of developmental research.* Dordrecht, Netherlands: Kluwer.

Thompson, A., & Thompson, P. (1996). Talking about rates conceptually, part II: Mathematical knowledge for teaching. *Journal for Research in Mathematics Education, 27*, 2–24.

Treffers, A. (1987). *Three dimensions: A model of goal and theory description in mathematics instruction—The Wiskobas Project.* Dordrecht, Netherlands: Reidel.

Treffers, A. (1991). (Meeting) Innumeracy at primary school. *Educational Studies in Mathematics, 22*(4), 309–332.

Whitney, H. (1988). *Mathematical reasoning, early grades.* Unpublished manuscript, Princeton University, New Jersey.

Chapter 5

Coordinating Social and Individual Analyses: Learning as Participation in Mathematical Practices

Michelle Stephan
Purdue University Calumet
Paul Cobb
Vanderbilt University
Koeno Gravemeijer
Freudenthal Institute

Traditionally, analyses of students' learning has been cast as an individual endeavor (cf. Piaget, Inhelder, & Szeminska, 1960), with social aspects, if accounted for at all, seen as catalysts for thought processes. Recently mathematics education research has seen a shift in emphasis from treating learning as solely an individual process toward analyzing learning in the social context of communities. As can happen in these kinds of profound shifts, a dichotomy was produced: that of subscribing to either an individual or a social perspective (see Cobb, 1994). Some, however, have attempted to transcend these two extreme positions by coordinating the two perspectives (Bauersfeld, 1988; Cobb, 1994). In their view, learning is simultaneously a social and an individual process, with neither taking primacy over the other. The primary goal of this chapter is to use the results of the classroom teaching experiment on linear measurement to detail an analysis of students' learning from a coordinated perspective. We will show with this sample analysis how students' learning can be characterized as having both social and individual aspects simultaneously. To this end, we view the forthcoming analysis as situating students' learning in the social context of their classroom community. For us, students' learning is seen as participation in the local, emerging mathematical practices. This view will become clearer as the sample analysis is presented.

The sample analysis found in this chapter gives an account of the collective learning accompanying the measurement sequence and thus serves as the backbone for revisions to the hypothetical learning trajectory laid out in chapter 4.

Although the main objective of this particular chapter is to discuss the theoretical notion of coordinating social and individual aspects of learning—the first of the three themes of the monograph—the sample analysis will illustrate each of the other two monograph themes, as well. The analysis will show that the use of taken-as-shared tools was integral to the evolution of the measuring practices. More theoretically, the analysis in this chapter also takes its place in the Design Research Cycle because it is an instance of classroom-based research and because it provides a rationale for revisions and continuation of the Cycle. We leave the tool-use theme implicit for now and take up its role in learning in chapter 6. We visit the third theme of this monograph—the Design Research Cycle—more specifically in the next chapter, as well.

This chapter is organized as follows. First, we present an analysis that coordinates the evolution of both the collective learning and two students' personal mathematical development. We document the emergence of five classroom mathematical practices and situate two students' learning within these practices. Second, we use the analysis as a context within which to discuss theoretical aspects of analyzing learning in social context. Specially, we argue that the mathematical practice analysis and complementary case studies serve as a more complete picture of the learning of the first-grade classroom. Further, we contrast this coordination with traditional cognitive and sociocultural analyses. Finally, we note the usefulness of the classroom mathematical practice as a construct for documenting collective learning and supporting learning in other socially situated classrooms.

THE EVOLUTION OF THE CLASSROOM MATHEMATICAL PRACTICES

This measurement sequence is divided into five phases, each corresponding to the emergence of a classroom mathematical practice. In this section, we discuss each phase of the measurement sequence, simultaneously elaborating each of the five mathematical practices. Concurrently, we discuss two students' individual ways of reasoning as they participated in, and contributed to, the emergence of each of the five mathematical practices. The two students chosen for this analysis had been targeted at the outset of the teaching experiment and were followed almost daily during the entirety of the project. These students were targeted on the basis of their distinct ways of reasoning at the outset.

Phase 1—Measuring Items in the Kingdom (3 Days)

At the beginning of the measurement sequence, the teacher told a story about a king who lived in a kingdom and sometimes needed to measure for various purposes. The king decided he wanted to use his foot to measure. The teacher asked the students to imagine that they were servants in the kingdom and to help the king determine how he might use his feet to find how long various items were. They decided that he could pace by placing his feet heel-to-toe with no spaces in between paces. During whole-class discussion, a student paced along the edge of a rug at the front of the classroom to illustrate how the king might pace. As she paced, she counted each step to keep track of how many steps had been taken. In subsequent instructional activities, students pretended to be the king and found how long various items around the classroom—the kingdom—were, using their own feet. Organized in pairs, each student took turns being the king and recorded what they found each time they measured. The classroom mathematical practice that became established as students participated in these activities involved measuring by covering distance. (Unlike Piaget et al. (1960), we do not distinguish between distance and length. For easier reading we use the word *distance* to refer to both the empty space between two points and the length of an object. We write the word *space* in our analysis when the students use the term themselves.)

The emergence of the first mathematical practice—Measuring by covering distance

The first mathematical practice that emerged involved a qualitative shift from measuring by counting the number of units used in the activity to measuring by covering distance. In addition, students had a variety of ways of participating in this measuring practice. Here we discuss the collective learning of the classroom community by documenting the establishment of the first mathematical practice and the learning of two students as they participated in, and contributed to, the emergence of the practice. First, we describe the negotiation process that led to the emergence and eventual establishment of the first measuring practice. Second, we detail the ways in which two students, Nancy and Meagan, participated in this practice. Third, we provide another snapshot of classroom activity in which the first measuring practice was negotiated.

Negotiating the purpose of measuring. On the day the king's-foot scenario was introduced, the teacher and the students discussed that the king could use his feet to pace the length of items in the kingdom. As students worked in pairs, we observed two distinct ways of counting paces. Some students placed their first foot down such that their heel was aligned with the beginning of the carpet and counted "one" with the placement of the next foot (see (a) in Figure 5.1). Other students placed their first foot such that the heel was aligned with the beginning of the rug and counted "two" with the placement of their next foot (see (b) in Figure 5.1).

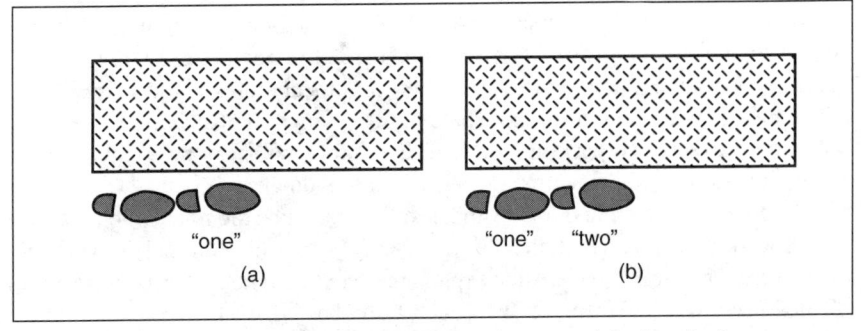

Figure 5.1. Two methods of counting as students paced the length of a rug.

We conjectured that those students who counted paces in the first manner (situation (a) in Figure 5.1) did not see their initial pace as covering a distance. In other words, for these students, the goal of pacing was to count the number of steps required to reach the end of the rug rather than to cover a length. We thought that if students contrasted these two ways of counting paces in the subsequent whole-class discussion, then the mathematically significant issue of measuring as covering distance might emerge as a topic of conversation. In the following whole-class discussion, the teacher and the students negotiated both the method and purpose of pacing. The teacher asked Sandra and Alice to show the class how each of them would measure the rug at the front of the classroom so that the students could contrast the two methods of counting paces.

T: I was also really watching how a couple of you were measuring. Who wants to show us how you'd start off measuring, how you'd think about it?

Sandra: Well, I started right here [*places the heel of her first foot at the beginning of the rug*] and went one [*starts counting with the placement of her second foot as in figure 5.1(a)*] two, three, four, five, six, seven, eight.

T: Were people looking at how she did it? Did you see how she started? Who thinks they started a different way? Or did everybody start like Sandra did? Alice, did you start a different way or the way she did it?

Alice: Well, when I started, I counted right here [*places the heel of her first foot at the beginning of the rug and counts it as one as in figure 5.1(b)*], "One, two, three."

T: Why is that different to what she did?

Alice: She put her foot right here [*places it next to rug*] and went, "One [*counts one as she places her second foot*], two, three, four, five."

T: How many people understand? Alice says that what she did and what Sandra did was different. How many people think they understand? Do you think you agree they've got different ways?

As this episode progressed, the two students, with the teacher's aid, articulated their initial methods and interpretations of measuring the length of the rug. In Sandra's first explanation, she indicated that she started measuring by placing her first foot down and counting the placement of her next foot as "one." Using Toulmin's (1969) model of argumentation, which is a useful analytic tool because the students did not view their contributions as constituting parts of this model, we argue that she was providing a claim—that is, that the rug to a certain point was eight paces long—and the data consisted of her method of counting—that is, counting "one" with her second step. In this bit of argumentation, Sandra had not made her warrant explicit to the class. As the argumentation continued, however, Alice presented an alternative explanation, after which the teacher attempted to make the warrants and backings explicit. In contrast with Sandra's explanation, Alice counted "one" the first time she placed her foot down and counted "two" with the placement of the second foot. Similar to Sandra, Alice did not explicitly articulate a warrant. As we see it, the question "Why is that different to [sic]what she did?" was an attempt to have Alice explain more explicitly how her interpretation of measuring led her to make the conclusion she did. In the next line, Alice contrasted the two solution methods by presenting Sandra's explanation. In

Toulmin's terms, Alice was providing a warrant, or information that described how each of those students came to their alternative solutions. Notice that Alice's warrant here contained exactly the same content as Sandra's explanation; however, we characterize it as a warrant, not data, because it serves the function of answering a challenge (see chapter 3 of this monograph). The teacher challenged Alice to justify why the two interpretations were different, how each student's data led to the contrasting conclusions that were drawn. In response to the challenge, Alice offered Sandra's explanation implicitly as a contrast with hers. In the next episode, we will see instances in which students provide more detailed warrants for each of Sandra and Alice's solutions.

Before we proceed to the next piece of dialogue, we note that, to this point, students' warrants for their measuring generally consisted of explaining their method of pacing, that is, which paces were counted. For students who paced Sandra's way, measuring initially appeared to signify going through the act of pacing without considering that their first foot covered a distance. To initiate a discussion in which the two different warrants were made more explicit, the teacher asked Melanie to begin pacing the length of the rug so that a piece of masking tape could be placed at the beginning and end of each pace. Once this record was made, the students who counted their paces Alice's way began to argue that Sandra's method of counting would lead to a smaller result because she missed the first foot. In the excerpt that follows, Melanie differentiated between the two ways of counting paces, whereas other students justified their particular method.

Melanie: Sandra didn't count this one [*puts foot in first taped space*]; she just put it down and then she started counting, "One, two." She didn't count this one, though [*points to the space between the first two pieces of tape*].

T: So she would count, "One, two." [*refers to the first three spaces, since the first space is not being counted by Sandra*]. How would Alice count those?

Melanie: Alice counted them, "One, two, three."

T: So for Alice, there's one, two, three there, and for Sandra, there's one, two.

Melanie: Because Alice counted this one [*points to the first taped space*] and Sandra didn't, but if Sandra would have counted it, Alice would have counted three and Sandra would have too. But Sandra didn't count this one, so Sandra has one less than her.

T: What do you think about those two different ways, Sandra, Alice, or anybody else? Does it matter? Or can we do it either way? Hilary?

Hilary: You can do it Alice's way or you can do it Sandra's way.

T: And it won't make any difference?

Hilary: Yeah, well, they're different. But it won't make any difference because they're still measuring, but just a different way, and they're still using their feet. Sandra's leaving the first one out and starting with the second one, but Alice does the second one and Sandra's just calling it the first.

Phil: They're both different ways. I thought Sandra's way would go higher than Alice's. Cause Alice started by ones and got three and Sandra only got two. Sandra would go higher cause she was lesser than Alice. . . .

Phil: She's fifteen [*refers to the total number of feet Sandra counted when she paced*]. Alice went to the end of the carpet [*he means the beginning of the carpet*]. Sandra started after the carpet. Hers is lesser 'cause there's lesser more carpet. Alice started here, and there's more carpet. It's the same way, but she's ending up with a lesser number than everybody else.

Alex: She's missing one right there. She's missing this one right here [*points to the first taped space*]. She's going "one," but this should be "one" 'cause you're missing a foot, so it would be shorter.

T: So he thinks that's really important. What do other people think?

Alex: Since you leave a spot, it's gonna be a little bit less carpet.

Melanie's first and second contribution provided a contrast between Sandra and Alice's measuring interpretations. In other words, Melanie was explicitly articulating, according to Toulmin, the warrants for each of the students' arguments. She described how the data led each student to make the conclusion that they did: "Because Alice counted this one and Sandra didn't...." The argumentation takes a turn when the teacher asks whether it made a difference which way one measured. In our view, the teacher was attempting to elicit a backing for the argumentations just presented. In other words, the teacher was implicitly asking which of these interpretations is mathematically acceptable to our classroom community. In the ensuing argumentation, we saw a lengthy negotiation in which a particular backing becomes taken-as-shared. By the end of the argumentation, Alice's interpretation was treated as legitimate in the public discourse, and the subsequent backings reflected the interpretation that measuring was about covering distance (e.g., Alex: "Since you leave a spot, it's gonna be a little bit less carpet"). If one measured Sandra's way, "you're missing a foot, so it would be shorter" [not as much carpet was measured].

This excerpt is significant because students' justifications suggest different interpretations of measuring. Initially, in such justifications as Melanie's, students were comparing two different sequences of paces instead of discussing that an amount of carpet was being measured. When the teacher asked the class if it mattered which way students began counting, the backings became articulated and involved measuring as covering distance. On the one hand, Hilary's explanation indicated that both Sandra and Alice were measuring the extent of the carpet but were counting their paces differently. For Hilary, measuring was a matter of convention, that is, students could count their paces either way even though they were different. On the other hand, Phil argued that the difference in the counting methods was indeed significant because Sandra would obtain a smaller number of feet: "there's more carpet." Alex explained that because Sandra was not counting her first foot, an amount of carpet was not being measured.

Alex's and Phil's justifications appeared significant because no one counted paces using Sandra's method in the whole-class discussion the following day. In fact, from this point on, the goal of measuring as covering amounts of distance was now beyond justification. In terms of Toulmin's (1969) model of argumentation, backings that involved measuring as covering distance dropped out of students' arguments, which is to say that measuring as covering distance became taken-as-shared. The foregoing episode is an illustration of the negotiation process involved in the constitution of a taken-as-shared mathematical practice. Documenting instances in public discourse in which backings emerge from the students to validate an argument is therefore helpful in delineating the emergence of new mathematical practices. As we see it, mathematical practices emerge during argumentations in which the participants provide new backings that shift the mathematical interpretations of the community to a new level. When backings for a particular interpretation drop out of the discussions or when alternative backings are contributed by a student and rejected by the community, we say that a mathematical practice has become established.

Many students, interestingly, had an intuition that pacing with smaller feet would result in more paces. Also, students were very careful not to leave any space in between their paces. The students' explanations of these two issues suggest that initially measuring was tied to the bodily action of pacing. If a student missed some space between her feet as she paced, a taken-as-shared explanation was that she had taken bigger steps, which result in fewer paces. If two students paced the same item and attained different results, it was argued that the person with larger-sized feet measured faster than the other person. In other words, her feet were bigger so she would take fewer paces and do less counting rather than cover more distance by each pace. These types of explanations further support the argument that the idea had become taken-as-shared that measuring was tied to the bodily act of pacing.

Meagan and Nancy. The episode discussed previously focused on the development of a taken-as-shared understanding of the purpose of pacing, that is, to cover distance. Nancy's participation in the taken-as-shared interpretation of pacing involved covering amounts of distance. For example, the day after Alice and Sandra's methods were contrasted, we observed Nancy measuring the length of a rug by counting her paces as Alice had done. When she reached the end of the rug, she said, "Forty-three and a little space." The fact that Nancy described the result of pacing as "Forty-three and a little space" suggests that she interpreted her pacing activity as covering an amount of distance. Our interpretation of Nancy's activity in this example is consistent with subsequent observations of her measuring activity. Such an interpretation was made possible by her participation in the whole-class discussion described in the previous section.

In contrast, Meagan appeared to participate in this practice by counting each pace as she walked the length of an item, without regarding each pace as covering distance or a part of the rug. For example, on the third day of the instructional sequence, Meagan measured the length of a wall as she worked with Nancy during small-group work. She began by placing the heel of her foot at one end of the wall and counted that foot as "one." She continued pacing heel-to-toe the entire length of the wall as she counted her paces. When she reached the end, she counted her last pace as "fifteen" even though part of the pace extended past the end of the wall. In this instance, because she saw no conflict in having part of her last pace extend past the end of the wall, Meagan seemed to simply count the number of paces she needed to take to reach the end of the wall. In other words, Meagan did not interpret her paces as distance-covering units, as segmenting the physical extent of the wall. Rather, her way of participating in the practice of measuring by covering distance involved the act of placing each foot and counting the number of steps that she took as she walked along the length of the wall. "Fifteen," for her, seemed to signify the number of paces she took her to reach the end of the wall rather than the amount of distance (wall) she had covered.

Our description of the second phase of the instructional sequence shows clearly that as Meagan participated in the first mathematical practice, she eventually made a conceptual reorganization concerning the underlying purpose of measuring, that is, to cover distance.

More learning in the first mathematical practice. Like Meagan, other students also had interesting ways of counting their last pace when part of it extended past the end of the item they were measuring. In fact, the teacher capitalized particularly on Perry's contribution, and later on Meagan's, in the following whole-class discussion. In the episode below, Sandra had paced the extent of a rug at the front of the room and found that it was 18 of her feet long. Perry argued that because part of Sandra's last pace extended past the end of the rug, she could not know how long the rug was.

Perry: My question was, What could it be, 'cause her foot is right here? [*He points to Sandra's last foot that extends past the end of the rug.*]
T: What's the problem with her foot being right there?
Perry: 'Cause then we can't figure out how many feet it would be.
T: Why not?
Perry: Well, because it's longer out.

...

[Note: An ellipsis such as this between lines of transcript indicates that irrelevant conversation has not been reported.]

Perry: Well you see, why would it be out here if that's not how long the rug is?
Max: Her foot's longer than the carpet.
Perry: That still doesn't matter. 'Cause it still doesn't make the feet different. The carpet can't move.
T: What do you mean the carpet can't move?
Perry: I mean the carpet isn't alive, so it can't move. So it has to stay right there, so it doesn't make a difference with the feet . . . and that doesn't tell us how many feet the carpet is . . . it doesn't count, they're still off the rug, but that doesn't tell us how long the rug is and that's what we're trying to figure out. So how can we know it if people's feet are going off the rug?
T: Is what you mean that they don't fit at the end exactly?
Perry: [*Agrees.*]

For Perry, an item could not be measured if a person's last foot did not fit exactly within the physical boundaries of the item. Perry understood that when a student's last foot extended past the endpoint, the result of pacing did not determine how many feet long the rug was. His backing for this interpretation consisted of the idea that the "carpet can't move." We interpret that statement to mean that as students paced, they were, in effect, defining the distance *as they measured*. For example, Samantha's last pace defined a distance that extended past the end of the rug. This newly defined distance did not fit with what Perry had anticipated. In other words, Perry anticipated that paces would fit exactly within the physical endpoints of the rug. In his view, when the last pace was shorter or longer than the endpoint of the rug, the rug could not be measured. To him, the rug itself seemed to take precedence over the measuring activity, and he could not conform his measuring activity to fit with the length of the rug; that is, as he said, "the carpet can't move," and the possibility of adjusting his measuring activity was not available for him. No one challenged Perry's backing, but they tried instead to develop ways to measure so that no part of their paces extended past the end of the item being measured. This lack of challenge indicates to us that Perry's interpretation was taken-as-shared in this mathematical practice.

To correct for their last pace extending past the endpoint of the carpet, many students simply turned their feet sideways so that the last foot fit exactly within the end of the rug, and they counted the whole foot as "one." The teacher asked whether counting the last pace in this manner was a good way to measure the extra distance at the end of the rug. Some students argued that they could not count the last pace as a whole foot because a whole pace was not needed. Thus, these students argued that the remaining distance was about a half a foot. The teacher remembered that Meagan had counted the turned pace as "one" in small group work earlier in the class period and asked students whether they could solve it that way. That solution method was immediately rejected by the class, indicating that measuring as covering distance was taken-as-shared. Finally, instead of turning the last foot and fitting it inside the remaining distance, some students did not place their foot at all and just visually estimated, for example, "Fourteen and a little space."

To make a clarification, we see the contrasting reasoning of Meagan and others as acts of participation in the first mathematical practice. Further, we consider those acts of participation as simultaneously contributing to the emergence of the mathematical practice. For example, when the teacher offered Meagan's reasoning as a possible solution method, the students rejected it. In fact, the rejection of Megan's solution by her classmates indicates that her act of participation contributed to the negotiation of the first mathematical practice. We view the coordination of social and individual perspectives in that sense: on the one hand, the first measuring practice emerged as Meagan and other students reasoned and contributed to it; on the other hand, as we will see in the next phase of the instructional sequence, Meagan and others reorganized their evolving interpretations as they participated in the evolving mathematical practice.

Describing the result of measuring. As students continued to engage in pacing activities, the idea that the result of pacing signified a sequence of steps that could be counted also became taken-as-shared. For example, if students paced the carpet and obtained a result of 15, "13" was taken-as-shared to be the distance covered by the 13th step rather than the whole span of the rug. As Hilary explained in whole-class discussion, the 14th, the 13th, and so on, paces were "what got you up to 15." For us, this way of describing the results of pacing could lead to a mathematically significant whole-class discussion. Recall that we were trying to support students' coming to act in a spatial environment in which measuring signified an accumulation of distances. Hilary's explanation, together with the fact that the community accepted her explanation, indicated that it had become taken-as-shared that the result of pacing signified a chain or sequence of individual paces. In other words, "fourteen" signified the distance covered by the 14th pace in the sequence rather than an accumulation of the distance covered by 14 paces.

In summary, the first mathematical practice emerged as students compared different ways of counting their paces. That measuring involved defining a distance *as* one measured appeared to be taken-as-shared. Defining this distance seemed to be tied to the bodily activity of pacing such that the physical placement of each pace defined the distance being measured. Also, the results of pacing signified a sequence of paces that could be counted to find how long an item was. Further, the last number word that was spoken signified the distance covered by the last pace rather than the accumulation of distances covered by all paces. Meagan and Nancy appeared to participate in this first mathematical practice in two different ways. Meagan's participation involved interpreting the activity of measuring as counting the number of paces she took to reach the end of the object; pacing did not necessarily signify covering distance at this point for her. Alternatively, measuring signified covering distance for Nancy.

Phase Two—Footstrip (7 days)

The students engaged in pacing activities for 3 days. On the 4th day, the teacher introduced a new scenario about smurfs who lived in a village and sometimes needed to know how long things were. The teacher explained that to find out how long things were, the smurfs used their food cans (Unifix cubes) to measure by placing them end to end and counting them. Our intent was for the collective activity of measuring with the Unifix cubes to serve as a basis for a more sophisticated tool to be developed later in the sequence. As students engaged in initial activities with the Unifix cubes, however, we learned that these new instructional activities were inappropriate for them. For example, one activity the teacher posed involved using food cans—Unifix cubes—to show the height of a wall, which was 41 cans tall, so that the smurfs could build a rope ladder tall enough to climb over the wall. As students attempted to solve this task, we soon realized that using the cans as a unit with which to measure was not taken-as-shared. Most students argued that it was impossible for the smurfs to find the height of the ladder unless they had enough cans to scale the wall on both sides. They imagined the smurfs actually climbing a column of cans to get to the top of the wall. The smurfs, therefore, needed enough cans to climb up the wall and enough extra cans to get down the other side. The difference between this new situation and the king's-foot scenario was in the contrast between interpreting the *results* of measuring and engaging in the *activity* of measuring. In the prior pacing activities, students *created* the measure of an object as they paced; students were engaged in the *activity* of measuring and actually carried it out. In contrast, with the Unifix cubes, students were asked to take the results of measuring as given, that is, the height of the wall was given to them instead of their having to create it by actually measuring. Our conjecture was that students needed to have more experience shortening their measuring activity and needed to engage in discussions that centered on what the *results* of measuring signified.

The emergence of the second mathematical practice—Partitioning distance with a collection of units

Creating a record of pacing. Given these reflections, the teacher returned to the king's-foot scenario the following day and told the students that the king was spending all his time measuring instead of doing things that were necessary to run a kingdom. She asked whether they could think of another way that the king could measure so he would not always have to measure things for other people. Students made several suggestions, including sending one pair of the king's shoes out to those who needed them and distributing several pairs of the king's shoes around the kingdom. The teacher then related that one of the king's advisers suggested putting a picture of the king's foot on a piece of paper. The students decided that they would need at least two feet so that a person who was using this picture would not skip any distance between two feet. The teacher asked students to draw five feet on a strip of paper and then asked how the king might use such a strip:

T: The king says he wants five. "I want five in a row." How could you use something like this?

Melanie: It would work because he would put the paper in front of itself.

Hilary: This would be faster. He'd have lots more feet, and if he had lots more feet, he could just take a big step with all the feet together and there were ten. It wouldn't be just ten, twenty. Each one would be the same size, and you could just add them all together.

In this excerpt, Hilary drew on her prior participation in the first mathematical practice to anticipate how using the footstrip would be more efficient. Instead of taking five little steps with their feet, students could now take one "big step" of five. More important, the students realized that solving problems would now be faster, indicating that some of them anticipated counting their individual paces more efficiently, that is, they would count faster by five paces than by single paces. This phenomenon of drawing on prior participation is exactly what we hope for when designing instruction for students. The instruction in this teaching experiment was designed to build on students' prior understanding so that when students encountered new problems, they would have a way to act. So, having come to this conclusion, each pair of students then created their own strips of five paces, which the class subsequently named *footstrips*.

Meagan. Meagan's way of participating in the first mathematical practice appeared to involve counting the number of paces she took to reach the end of an item, without necessarily regarding her measuring activity as covering the physical extension of an item. As she participated in the whole-class discussion during the constitution of the first practice, Meagan appeared to reorganize the goal of measuring such that paces now became distance-covering items. Evidence for this result comes as Meagan participated in a whole-class discussion on day 5 in which two groups of students were comparing the footstrips they had made. One pair of students had constructed the footstrip seen in Figure 5.2.

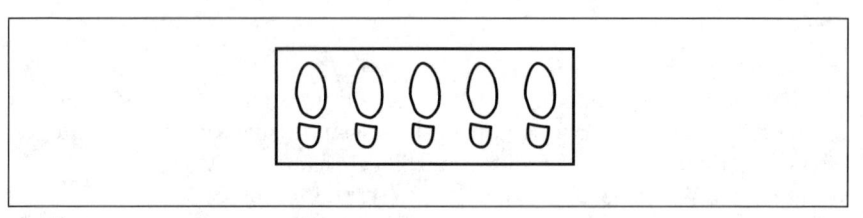

Figure 5.2. One pair of students' footstrip.

Meagan argued that if they used their footstrip "and you were measuring the rug, you would be missing some spaces." Paces, or records of paces, now seemed to be distance-covering entities for Meagan because she argued that the paces were arranged on the students' footstrip in such a way that measuring with that particular footstrip would allow amounts of distance to be missed when iterating the footstrip. In fact, she and Nancy had constructed their footstrip by drawing five paces heel-to-toe and marking a line and numeral after each pace (see Figure 5.3). Meagan explained that when they were making their footstrip, they had placed numerals at the end of each pace so they would not miss any distance in between paces.

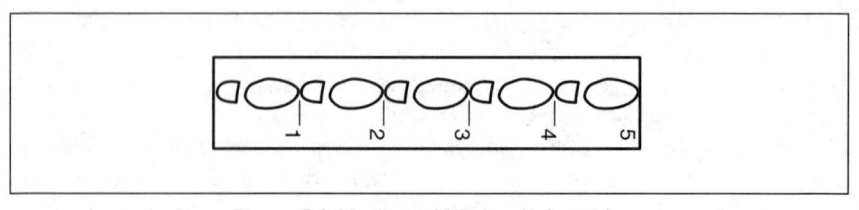

Figure 5.3. Nancy and Meagan's footstrip.

Nancy and Meagan—Measuring with the footstrip. The 7th day marked the first instance in which students measured with their footstrips. They were asked to work in pairs and measure items on the playground. Observations of Nancy and Meagan's activity indicated that measuring, for them, was dependent on the act of placing down the footstrip. For example, consider how Meagan and Nancy measured the length of a small shed, which actually measured 24 1/2 paces. They placed one end of the footstrip at the beginning of the shed and counted "five." Next they iterated the footstrip end to end, counting by fives as they did so, "Five, ten, fifteen, twenty, twenty-five, twenty-five and a half [sic]" (see Figure 5.4).

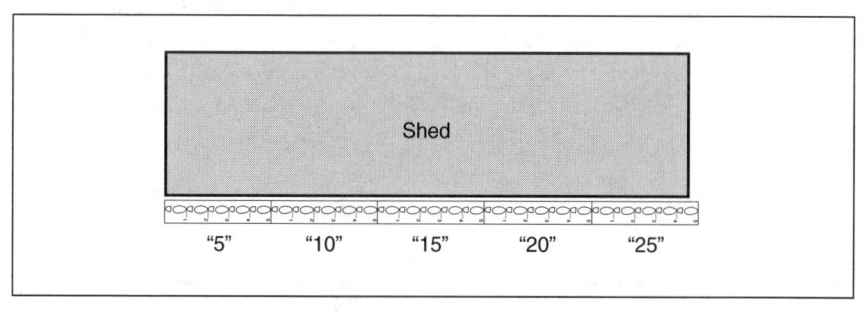

Figure 5.4. Meagan and Nancy's method of measuring a shed.

Instead of placing the footstrip a fifth time and counting up by ones from 20, that is, 21, 22, 23, 24 1/2, Meagan and Nancy placed the footstrip, counted 25, and then counted the half a foot that extended beyond the end of the shed. When they began measuring the shed, they seemingly intended to cover the spatial extension of the shed with an exact number of footstrips. The final placement of the footstrip, however, extended past the end of the shed. This result did not fit with what they had originally intended to do, which was to cover the spatial extent of the shed with a sequence of footstrips. Thus, they placed the footstrip a fifth time, saying "twenty-five," where 25 signified the last part of the spatial extension of the shed. They then realized that the footstrip extended past the end of the shed and added an extra half a pace. Measuring seemed to have become inseparable from the activity of iterating the footstrip. Because the distance they had measured, the distance covered by 25 paces, did not fit with what they had intended to cover, they added the extra half pace.

This interpretation of measuring with the footstrip was typical of both Nancy's and Meagan's activity initially. As they participated in the negotiation of a second mathematical practice described below, both eventually reorganized their understanding.

Constitution of the second mathematical practice. In the whole-class discussion that follows, the taken-as-shared interpretation of measuring evolved and the second mathematical practice emerged. Whereas Nancy reorganized her understanding in this episode, Meagan did not. A fundamental issue reemerged as students participated in the whole-class discussion following the small group work described in the foregoing. Recall that the previous mathematical practice involved covering the spatial extension of items with an exact number of paces and that this activity was integrally tied to bodily action. As the second mathematical practice was constituted, the idea that part of a unit of measure can extend past the endpoint of an item became an explicit focus of conversation. During the whole-class discussion that follows, Porter and Sandra used their footstrip to find the length of a cabinet that was located at the side of the classroom and ended against a wall. After three iterations, Porter and Sandra found that they did not have enough room to place another full footstrip. Instead of sliding one end of the footstrip up the wall, Porter and Sandra placed the farthest end of the footstrip against the wall so that it overlapped the third placement of the footstrip (see Figure 5.5).

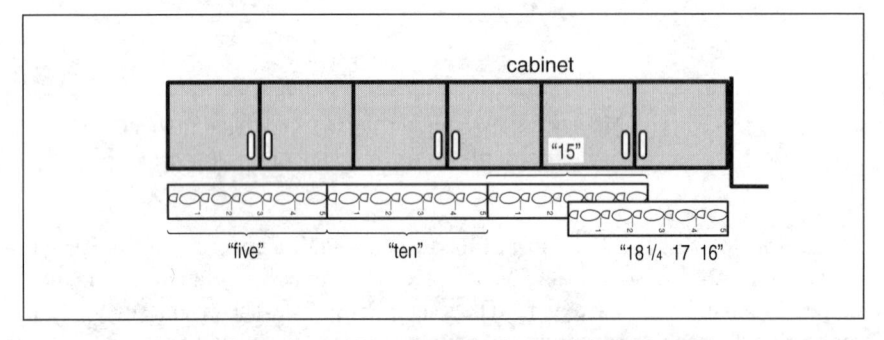

Figure 5.5. Sandra and Porter's method of measuring the cabinet.

Sandra and Porter then counted "Sixteen, seventeen, eighteen and a fourth" back along the footstrip from the wall until they reached the end of the third placement. Because many students seemed to be confused by this solution method, the teacher suggested that it might be easier for others to understand if Porter and Sandra moved the footstrip so that the excess ran up the wall. Lloyd argued that moving the footstrip up the wall would not work.

Lloyd: Uh-uh. It doesn't measure the whole cabinet. But it won't work.
T: So Lloyd says it won't work. That's interesting. What do others think?
Pat: I think it would be like, uh, I think that would be just 18 from measuring the cabinets.
Lloyd: I don't think so. You're supposed to be measuring from down there [*points to the beginning of the cabinet*] to right here [*points to where cabinet meets the wall*]. I mean to the wall. And that is measuring up the wall. It's supposed to stay here.
Pat: But, well, it stopped at the end of the cabinet.
Lloyd: It doesn't work. 'Cause she said from that box [*he is pointing to a box that sits at the beginning of the cabinet*] all the way to the wall, and it's measuring up the wall. She said from the box to this wall right here [*points to where the wall and cabinet meet*].

T:	We've got two people saying different things here.
Sandra:	I have a question for Pat. She didn't say go all the way up to the wall.
Pat:	Yeah, but if you just left it right here [*with footstrip extending up the wall*], then you wouldn't have these extra feet right here [*pointing to the 2 3/4 feet extending up the wall*].
T:	So, Pat. Are you kind of in your mind seeing where the wall, the wall . . .?
Lloyd:	I see something [*excitedly*]. The half and the two feet. Fifteen, sixteen, seventeen, and a half foot [*points to each foot in the last footstrip as he counts*]. It's about this. I think he meant to, like, pretend that that part was cut off right there [*points to part of strip going up the wall*].
Pat:	Yeah.
T:	So you just imagine that in your mind, Lloyd? Does that help, Edward?
Edward:	I agree.
Max:	Some people could think that he might be measuring up the wall, like to get to twenty.
Lloyd:	No. That's not what he's doing. He's just cutting it off. I think we solved the problem.

In this episode, Lloyd initially reasoned that because the footstrip extended past the endpoint of the cabinet, then this method of measuring would not work. In Toulmin's terms, Lloyd claimed that measuring in this manner did not work, the data being that the footstrip went past the edge of the item. For Lloyd and others, placing the footstrip defined the distance being measured. When Lloyd placed the footstrip down the fourth time, the distance being measured no longer stopped at the end of the cabinet but rather extended to the far end of the footstrip. Hence, measuring this way "would not work" because the distance that was being covered by the footstrip extended past the physical extension of the cabinet. In this way, measuring was tied to the activity of placing the footstrip. This measuring activity grew out of students' participation in the first mathematical practice, where measuring by pacing was tied to bodily action; the footstrip was a record of their physical actions.

As a result of Pat's challenge, Lloyd provided a warrant, saying, "They're supposed to be measuring" from one end to the other. In our view, he is drawing on his participation in the prior mathematical practice of covering distance with an exact number of units. This practice had been taken-as-shared, and, therefore, Lloyd did not produce a backing for his argument. In contrast, Pat argued that he could, in fact, place the footstrip so that part of it extended up the wall, and he would still only measure the cabinet. The data in this argument involved counting only those paces that were needed and ignoring the extra paces. Students still had difficulty with Pat's interpretation, so the teacher, for his part, attempted to elicit a backing for Pat's argument when he said, "Are you kind of in your mind seeing where the wall...?" The teacher was helping Pat articulate how he mentally structured the distance that was being measured. Lloyd reorganized his measuring understanding and recast Pat's justification in terms of pretending to cut off the footstrip at a certain point. Significantly, mentally cutting the footstrip along any point became a taken-as-shared backing in this episode. As subsequent whole-class discussions proceeded, other students used mentally cutting the footstrip as a backing to justify why they extended parts of a footstrip past the end of an object and counted only those paces that were relevant to their measuring. This shift in the content of students' justifications—the evolution of the backing—marks the emergence of a second mathematical practice involving partitioning distance with collections of units, in this case, with five paces. When mentally cutting the footstrip eventually dropped out of students' arguments, we claim that the second mathematical practice was relatively stable. The remaining mathematical practices also came under the same methodological scrutiny as the two practices we have described, although we do not present a detailed analyses of the discourse with warrants and backings because we do not want to overwhelm the reader.

In this second mathematical practice, the distance to be measured was independent of activity, independent of the placement of the footstrip. Now the distance to be measured took priority over the measurement activity instead of the footstrip-placing activity's defining what was being measured. Measuring was no longer tied to the physical act of placing a footstrip; rather, the footstrip's being mentally cut as the need arose became taken-as-shared.

The initial interpretation of not extending a unit past the farthest endpoint had become taken-as-shared in the first mathematical practice but was renegotiated during the establishment of this second mathematical practice. The explicit attention given to this issue in this phase of the sequence can be attributed to the taken-as-shared use of the footstrip. When measuring by pacing the length of an item, students could simply remove their last foot if it extended past the endpoint and refer to the extra distance by pointing to it. Now, when using the footstrip, three or four extra paces might extend past the endpoint, but these paces were physically inseparable from one another. Therefore, opportunities to discuss what happened when paces extend past the endpoint of an item were made possible.

Meagan and Nancy: Reorganizing their measuring activity. During the pivotal whole-class discussion described previously, Nancy, not Meagan, appeared to reorganize her prior measuring activity. For the remaining occasions in which Nancy measured with the footstrip, she coordinated measuring with a strip of five paces and counting single paces. In fact, Nancy continually challenged Meagan's measuring activity during small group and whole-class discussions in which she measured with her partner. For example, in the following episode, Meagan, for whom measuring depended on the placement of the footstrip, and Nancy were measuring string that signified the length of fish that the king had caught. The string actually measured 12 1/2 paces long, and Nancy explained her result:

Nancy: We didn't have enough room for fifteen, so I went ten, eleven, twelve, thirteen and a half [*points to each of the first three paces on the third iteration of the footstrip*].

For Nancy, measuring with the footstrip now appeared to signify a more efficient form of pacing; "fifteen" was the amount of distance covered by 15 single paces. Thus, "We didn't have enough room for fifteen [paces]" indicated that, for her, the string was not long enough to be covered by exactly one more footstrip. Hence, "ten" signified the distance covered by 10 paces, and then she counted "eleven, twelve, thirteen and a half." This approach illustrates that Nancy now coordinated measuring with the footstrip with the result she would obtain if she actually paced. Measuring no longer depended on the placement of the footstrip; the third footstrip did not define the distance she had measured thus far. Rather, Nancy mentally envisioned cutting the last footstrip on half of the 13th pace. This conception is a significant shift in her understanding from the beginning of the second phase of the measurement sequence.

In contrast, Meagan had difficulty coordinating the result of measuring with the footstrip with what she would have gotten if pacing. Meagan eventually, however, came to coordinate measuring with the footstrip and individual paces. She reorganized her thinking during whole-class discussions in which students' interpretations of the results of measuring were made explicit and during pair work with Nancy. Although both Meagan and Nancy made this coordination while measuring with the footstrip, the result of measuring signified different things for them. Meagan, on the one hand, interpreted the result of measuring as a sequence of individual paces, whereas Nancy understood the result as an accumulation of distance. For example, when a researcher asked Meagan to show on the footstrip how long something 3 1/2 feet would be, Meagan pointed to half of the fourth pace:

T: Where's three?
Meagan: There [*points to the third pace*].
T: So three is that one foot there?
Meagan: Yeah. Three and a half of the fourth foot.
T: Could three mean all of these feet [*points to the space between the beginning of the footstrip and the end of the third pace*]?
Meagan: That's number one [*points to the first pace*], that's number two [*points to the second pace*], and that's number three [*points to the third pace*].

From this exchange, Meagan appears to have interpreted the result of her measuring activity as a sequence of individual paces. "Three" signified the distance covered by the third pace rather than the distance covered by all three paces. When the researcher asked her to clarify if "three" could mean all three feet together, Meagan indicated clearly that "one" meant the first pace; "two," the second; and "three," the third pace. This interpretation is significant in that we were trying to support students' coming to act in a spatial environment in which measuring signified the accumulation of distances. As Meagan participated in the practice of partitioning distance with a collection of units—for us, the cognitive or mental activity of segmenting space with units—the result of measuring signified a sequence of individual paces for her. As can be seen in the episodes related here, the results of partitioning were different for Nancy and Meagan. Even though she did not interpret the result of measuring as an accumulation of distance, Megan's interpretation is clearly more sophisticated than how she initially participated in the practice.

Nancy, in contrast, interpreted the result of measuring as the accumulation of distance covered by paces. When asked to show how long something 8 1/2 feet would be, Nancy iterated the footstrip twice and replied:

Nancy: [*Sweeps her hand over the space from 0 to 8.*] Eight and a half [*points to half of the eighth foot*].

Nancy clearly indicates by the sweeping motion of her hand that she interpreted the result of measuring as an accumulation of distance. "Eight" signified the distance covered by 8 paces, as indicated when Nancy demonstrated that 8 would stretch from the end of the 8th pace to the beginning of the distance. In this episode, Nancy illustrated that the "half" from "eight and a half" is located on the half of the eighth foot. This answer is not in conflict with our claim that measuring for her signified an accumulation of distance. Rather, where to locate the half, for the teacher, was a matter of convention—either 8 and a half more or half of the eighth foot. The teacher, as a consequence of observing such solutions as Nancy's, established that the half is conventionally located on half of the next foot.

This kind of interpretation remained consistent throughout the rest of the measuring activities involving the footstrip. In fact, Nancy and Meagan sometimes measured during whole-class discussions. In this way, Nancy's and Meagan's participation in the practice of partitioning distance with a collection of units contributed to the constitution of this taken-as-shared way of measuring with the footstrip. In other words, as they and the class discussed contrasting interpretations, the understanding that the *placement* of the footstrip no longer defined the distance measured became taken-as-shared in the public discourse. Measuring became independent of the activity of iterating, and the footstrip could be mentally cut along any point as the need arose.

Conclusion of phase 2. In summary, the first mathematical practice involved measuring by covering distance, and the result of measuring signified a sequence of individual paces that extended to the end of the item being measured; measuring was tied to the bodily act of pacing. The second mathematical practice evolved from the first mathematical practice in that the footstrip was a record of their paces; hence, that similar issues emerged as topics of conversation was significant. Initially, measuring with the footstrip seemed to be integrally tied to the placement of the footstrip. As in the previous practice, the physical activity took precedence over the physical extension of an item. The idea became taken-as-shared, however, that the footstrip could be mentally cut; the extension of the item now took precedence over the activity. Measuring had to conform to the extension of the item rather than the other way around. Therefore, the second mathematical practice involved a collective reorganization of the first practice. In addition, a transition in the taken-as-shared argumentations about the result of measuring began. As illustrated by Nancy's reasoning, she and several students began talking about the result of measuring as a whole measured distance—an accumulation of distance—rather than a sequence of individual paces. Further, they used their hands to show the accumulation of distances measured, for instance, from 0 to 8, and referred to these distances as "the whole eight." This way of talking and gesturing became taken-as-shared later in the instructional sequence, but the transition seemed to begin as students measured with the footstrip.

Phase 3—Smurf Village (11 days)

In this phase of the instructional sequence, the teacher reintroduced the smurf-village scenario. Initially, students were given a bag of Unifix cubes and asked to find how long various items around the smurf village, the classroom, were. Subsequent instructional activities included giving students a piece of adding machine tape and telling them that it signified the length of an animal pen. They were also supplied several other pieces of adding machine tape that signified the lengths of different animals in the kingdom, such as a dog, a cat, or a horse. Students were then asked to determine whether each of the animals would fit in the pen. If the animal did not fit exactly, the student was asked to find how much longer or shorter than the pen the animal was. The taken-as-shared interpretation of measuring items in the classroom involved making a bar or rod of cubes that stretched the length of an item and then counting the cubes. The teacher suggested that the students might find a more efficient way of measuring with the food cans and asked students whether they could think of a way that the smurfs could measure without carrying around a whole bag of cubes. Students gave several suggestions that included carrying only a bar of 10 cubes, which they named a *smurf bar* because they had previously been told that smurfs were about 10 cans, or cubes, tall. Instructional activities with the smurf bar included having students find the lengths of various items around the room and cutting pieces of paper signifying different-sized wooden boards for building a smurf house. The instructional activities in this phase of the sequence differed from those in the first two phases in that now the students used a physical entity as a measuring tool instead of using a part of their body or a record thereof. The mathematical practice that emerged as students participated in the activities described in the foregoing involved measuring by accumulating distances.

On the 14th day of the instructional sequence, the class measured with a bar of 10 cubes for the first time. The teacher asked the students to work in pairs and to measure items in the classroom. Using a smurf bar for the first time, Meagan began measuring the height of an animal cage. She placed one end of the smurf bar at the bottom of the cage and said, "ten." She iterated the bar end to end along an edge of the cage and counted "ten, twenty." She placed the bar a third time and counted "thirty" even though the third iteration extended past the top of the cage. Then, she counted the cubes, saying, "thirty-one, thirty-two, thirty-three" (see Figure 5.6).

For Meagan, counting 30 as she placed the smurf bar for a third time seemed to mean that the cubes within that iteration should be counted "thirty-one, thirty-two, thirty-three . . . ." This way of reasoning indicated that Meagan was not coordinating measuring using the bar of 10 with measuring with the individual cubes

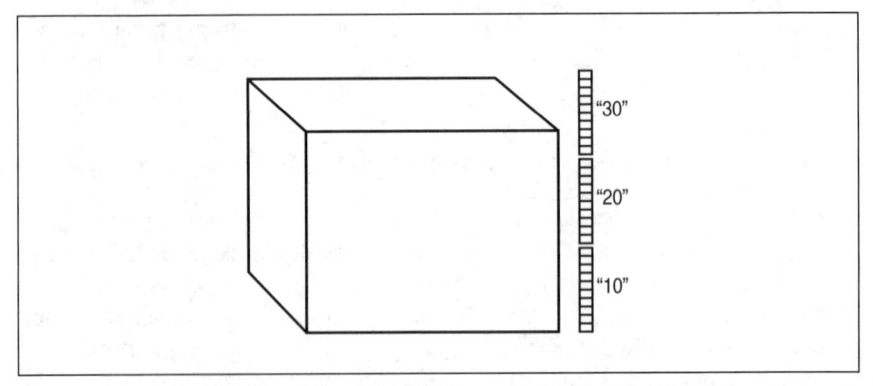

Figure 5.6. Meagan's method for measuring the height of a cage.

of which the bar was composed. In other words, iterating with the bar of 10 did not signify a record of measuring with individual cubes for her.

Nancy indicated that she disagreed with Meagan's measurement and remeasured the height of the cage by counting as follows: [iterates the bar once] "ten," [iterates the bar a second time] "eleven, twelve, thirteen, ..., twenty," [iterates the bar a third time] "twenty-one, twenty-two, twenty-three." For Nancy, iterating the bar of 10 appeared to signify the distance covered by cubes thus far, which can be seen by her counting by single cubes to justify her method of measuring. Although Meagan accepted Nancy's measurement, she continued to measure in the manner she had before. The fact that Meagan did not reorganize her understanding in this episode might be due to the calculational nature of Nancy's explanation. Nancy described only how she counted cubes; she did not specify her interpretation of distance that underpinned her measuring activity. Other students interpreted their measuring with the smurf bar in the same manner as Meagan; hence, the teacher decided to ask students to compare and contrast their differing interpretations in whole-class discussions. The teacher hoped that an accumulation-of-distance interpretation would become a topic for whole-class discussion. As we see in the next section, an accumulation-of-distance interpretation did become an explicit focus in subsequent public discourse.

The emergence of the third mathematical practice—Measuring by accumulating distance

The episode that follows illustrates a pivotal whole-class discussion in which an accumulation-of-distance interpretation was an explicit focus of conversation. This excerpt is significant because students' explanations indicate that an accumulation-of-distance interpretation became taken-as-shared during the discussion. Below, Alice and Chris were asked to cut a piece of adding machine tape that signified wooden boards to be used for a smurf house so that it measured 23 cans long. Alice iterated the smurf bar twice and uttered "ten, twenty" after each iteration. Next she placed the smurf bar down once more to find a length of 23 but picked it up, broke off three cubes from the smurf bar, and laid the three cubes at the end of the second iteration to indicate a length of 23:

Edward: I think it's thirty-three [*points to where they have marked 23 with the three cubes*] because ten [*iterates the smurf bar once*], twenty [*iterates the smurf bar a second time*], twenty-one, twenty-two, twenty-three [*counts the first, second, and third cube* within *the second iteration, thus measuring a length that was actually 13 cubes*].

...

T: Let's be sure all the smurfs can understand, 'cause we have what Alice had measured and what Edward had measured. We need to be sure everybody understands what each of them did. So Edward, why don't you go ahead and show what it is to measure twenty-three cans.

Edward: Ten [*iterates the smurf bar once*], twenty [*iterates the smurf bar again*]. I changed my mind. She's right.

T: What do you mean?

Edward: This would be twenty [*points to the end of the second iteration*].

T: What would be twenty?

Edward: This is twenty right here [*places one hand at the beginning of the "plank" and the other at the end of the second iteration*]. This is the twenty.

...

Then, if I move it up just three more. There. [*Breaks the bar to show 3 cans and places the 3 cans beyond 20*] That's twenty-three.

...

Phil: When you put the second down, that's the whole twenty [*points to the space from the beginning of what was the first iteration to the end of the second iteration*].

In the course of his explanation, Edward reconceptualized his measuring activity. Initially, for Edward, placing the smurf bar down for a second time defined the cubes in the second decade, the twenties. In other words, if he were to count the individual cubes, he would count "twenty, twenty-one, ..." Thus, although iterating involved a shortening of counting by individual cubes, it was based only on a number-word relation. "Twenty" was the number word associated with the second placement of the smurf bar rather than with the amount of distance covered by 20 cubes. After reflecting on Alice's explanation, however, Edward came to view "twenty" as signifying the length covered by 20 cubes and realized that 21, 22, and 23 must extend beyond the length whose measure was 20. Similar to Alice's reasoning, now if he were to count individual cubes in the second iteration, he would count "eleven, twelve, ..." For Alice, measuring with the smurf bar was a shortening of covering distance with individual cubes and involved an accumulation of distance. To clarify the meaning of the number 20, Phil explained that 20 signified the length from the beginning to the end of the second iteration, not the distance covered by the last 10 cans.

The main point of this excerpt is not only that Edward and others possibly reorganized their prior understanding but also that students negotiated a taken-as-shared interpretation of measuring with the smurf bar. When Alice initially explained how she measured 23, she iterated the bar twice and broke off three extra cubes to show 23. That she provided no backing for why this solution was acceptable is significant. Edward challenged her explanation, offering a justification in which he did not coordinate measuring with a bar and measuring with individual cans. As Edward reorganized his understanding in the midst of a rejustification, Phil explicated a backing that involved an accumulation of distance. As the whole-class discussion carried on, no student rejected Phil's explanation. In fact, other students used this way of reasoning as backing in subsequent conversations, with the backing eventually dropping out of students' arguments. Furthermore, when students offered an explanation that did not reflect an accumulation-of-distance interpretation, the community rejected it. We claim that in this way measuring by accumulating distance became taken-as-shared. That physically iterating along an item created a partitioned distance that could be structured in collections of tens and ones also became taken-as-shared.

Meagan and Nancy. Our observations of Nancy and Meagan's measuring activity for the remainder of this phase indicated that both girls interpreted the result of measuring as an accumulation of distance and coordinated measuring by 10 cans and individual cans. Meagan, however, was better able to coordinate measuring with tens and ones when she symbolized the results of her measuring activity. As she measured an object, she kept track of the number of cans by recording each iteration of 10 with masking tape or numerals. As Meagan reasoned with her record of iterating, two iterations signified the accumulation of distance covered by 20 cubes. Another way to state this outcome is that the number word "twenty" signified a numerical composite (Steffe, Cobb, & von Glasersfeld, 1988), an entity or a distance partitioned by 20 cubes rather than the distance covered by the 20th cube. This episode brings to the fore the role of symbolizing in Meagan's activity. As she reasoned with her record of iterating, Meagan interpreted the result of each iteration as an accumulation of distance covered by cans. Further, when reasoning with such symbols, she coordinated measuring by iterating a bar of 10 and measuring by iterating single cubes. Thus, as she participated in the prior conversations in which mathematically significant issues arose, she developed relatively powerful ways of reasoning with symbols. When she did reason with symbols, Meagan coordinated measuring with a bar of 10 and measuring by iterating single cubes. When Meagan did not reason with symbols, however, she did not make such a coordination.

Although Nancy symbolized her activity when she worked with Meagan, on her own, Nancy coordinated measuring with a bar of 10 and measuring with single cubes without symbolizing. Further, Meagan's reasoning with symbols was an act of participation in the emerging mathematical practice and constituted not only her learning but also a contribution to the taken-as-shared practices of the community.

Conclusion of phase 3. In summary, the emergence of the third mathematical practice of measuring by accumulating distances can be seen to grow out of the second mathematical practice. As a result of participating in the second mathematical practice of partitioning distance with collections of units, taking the result of measuring as a given had become taken-as-shared. As the third practice was established, a numeral, such as 23, signified an object's measure, and immediately iterating collections of 10 cubes to specify its length was beyond justification.

Phase Four—Measuring with the Measurement Strip (5 Days)

In this phase of the instructional sequence, the teacher built on the students' measuring activity to introduce a measurement strip that was 100 cans long. Students had engaged in instructional activities in which they cut pieces of adding machine tape that signified lengths of pieces of wood to be used for a smurf raft. For example, students were asked to cut pieces of wood that were 30 cans, 22 cans, and 5 smurf bars long. Whereas most students iterated a smurf bar and cut the paper at the end of the last iteration, Nancy and Meagan actually recorded each iteration of 10 on the pieces of paper (see Figure 5.7).

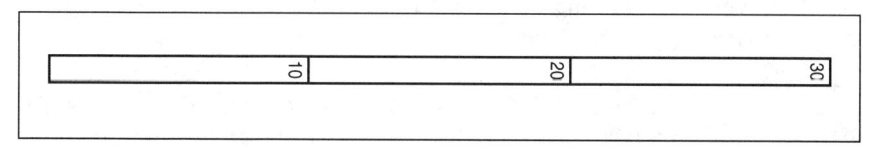

Figure 5.7. Nancy and Meagan's wooden board measuring 30 cans.

The teacher built on this innovation by suggesting to students that Nancy and Meagan's paper might be easier to carry around than a bag of cans. One student suggested they carry around a strip that was 10 cans long. The teacher responded that she liked their idea but that the smurfs decided they wanted to make a strip that was 50 cans long. The teacher asked the class to construct measurement strips that were 50 cans long and that did not require them to carry any cans with them. As we observed pairs of students, we noticed that most students constructed their strips as in Figure 5.8. An interesting feature of these strips was the fact that no individual lines were drawn to signify cubes within the decades. In fact, every pair of students constructed their 50-strips by iterating a smurf bar and marking only the end of the bar with a numeral until they had made a strip that was 50 cans long.

One interpretation of the students' constructions could be that each iteration of 10 that they marked on the strip did not signify composites of 10 but simply records of measuring with the smurf bar. Another interpretation, however, takes note of their participation in the previous mathematical practice involving iterating bars of 10. Consider how students measured the length of a table 43 cans long using

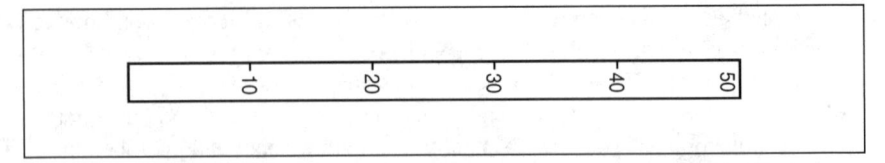

Figure 5.8. Initial constructions of the measurement strip

a smurf bar. Most students would iterate the smurf bar "ten, twenty, thirty, forty," iterate the bar one more time, and count the first three cubes in that iteration to obtain 43. "Forty" signified what students would get if they counted the cubes in the first four iterations by ones, and they needed only to count three individual cubes beyond "forty." Thus, in making the 50-strip, students marked only the end of each iteration because they seemed to have no need to know where the individual cubes ended when measuring something, say, 50 cans long. Hence, the construction of the 50-strip can be traced back to students' participation in the third mathematical practice of measuring by accumulating distances. The marks that students drew on the strip were a record of their activity of iterating the smurf bar where they had no need to record individual cubes.

The next day, the teacher asked students to make strips that were only 10 cans long instead of 50 cans. We thought that to support the emergence of a measurement strip 100 cans long, students might build on their constructions of 10-strips more naturally than 50-strips. The intent of the 10-strip was to build on the previous practice of measuring by accumulating distances; it fit with the activity of iterating to find a measure. At least two pairs of students cut paper 10 cans long and marked individual cubes on the strip. In a subsequent whole-class discussion, the need for individual markings arose from the students and every pair revised their strips.

In the absence of symbolizing, both Meagan and Nancy had difficulty coordinating measuring with a 10-strip and measuring with individual cans. Both of them, however, clearly understood the result of iterating as an accumulation of distance.

The emergence of the fourth mathematical practice—Measuring with a strip of 100

Subsequent instructional activities with the 10-strip included measuring items around the classroom. In a whole-class discussion on the 24th day, two students came to the white board at the front of the classroom and showed how they would measure it with their 10-strip. They iterated their strip, saying "ten, twenty, . . . ," and recorded the endpoint of each iteration with a marking pen (see Figure 5.9). They argued that counting by tens would be easier than counting by ones each time. Then the teacher taped seven 10-strips underneath their drawing and commented that each strip meant 10 cans they did not have to count by ones. In this way, the 100-can measurement strip emerged as a record of the activity of measuring. The teacher asked students such questions as "Where is twenty-five?" and "How long would something twenty-seven be?" During this discussion, one student illustrated that 49 cans would stretch from the beginning of the white board to the line signifying the end of the 49th cube. Nancy countered, however, that 49 actually extended from the beginning of the board to the line signifying the end of the 39th. To justify her way of reasoning, Nancy pointed to each of the first four spaces and counted, "ten, twenty, thirty, forty," and then, "forty-one, forty-two, ..., forty-nine" within the space signifying the fourth iteration of the 10-strip. Because the new measurement strip intentionally had its roots in iterating the 10-strip, a reasonable and desirable outcome is that students would draw on their participation in the prior practice associated with iterating a collection of ten to interpret their activity with the new tool. Other students countered her argument by starting with the numeral 20 and counting by ones up to 39 to show Nancy that where she pointed actually marked 39 cubes. Nancy accepted this justification and redescribed her solution method by counting by ones to show that what she originally thought was 49 was only 39 cans long.

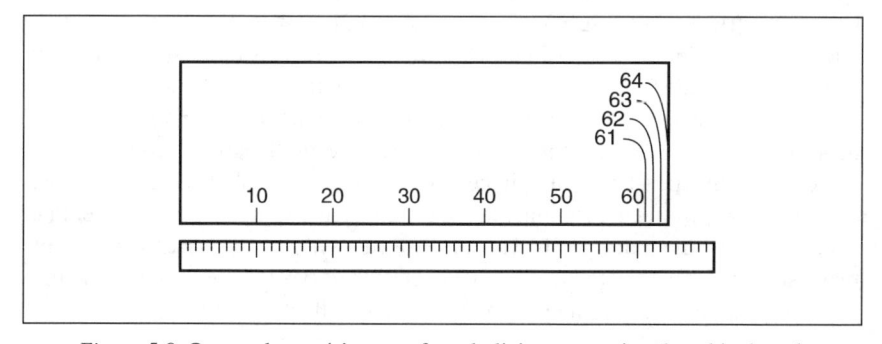

Figure 5.9. One student pair's way of symbolizing measuring the white board.

Thus, as students negotiated what became a taken-as-shared way of interpreting measuring with the strip of 100 in the foregoing example, their warrants consisted of counting individual cubes, and backings of accumulation of distance re-emerged in the context of reasoning with a new tool. The result of measuring with the 10-strip quickly came to signify the accumulation of distance as the community rejected Nancy's contribution. From this point on, no student during subsequent whole-class discussions had difficulty coordinating measuring with a 10-strip with measuring with individual cubes, indicating that it had become taken-as-shared. Such an interpretation was supported by the conceptual discussion illustrated previously in which students folded back to counting by individual cubes. To clarify, we view Nancy's contribution here as both her reasoning, and reorganization, and an act of contributing to the stability of the third mathematical practice. The community did not treat her act of participation as legitimate, indicating that the third mathematical practice had indeed been established. Her contribution also can be viewed as contributing to the negotiation of measuring with a new but similar tool.

Meagan and Nancy. As illustrated in the foregoing example, Nancy reorganized her understanding during the whole-class discussion. From this point on, Nancy and Meagan no longer had difficulty coordinating measuring with a 10-strip and measuring with individual cans. This reorganization was made as they participated in, and contributed to, the constitution of the fourth mathematical practice. To be sure, the reorganization was supported by the teacher's record of iterating that had played such an important role in each child's mathematical learning.

The measurement strip. After the discussion in which the teacher taped seven 10-strips on the whiteboard, she gave each pair of children a paper measurement strip that measured 100 cubes long (see Figure 5.10). Students engaged in various activities, such as using the measurement strip to find the lengths of items in the classroom. During activities such as these, the mathematical practice of measuring with a strip of 100 became established. Initially, many students measured by laying the strip down alongside the item and counting by tens and ones until they reached the endpoint of the item. Significantly, many students measured by "building" the extent of an item by iterating or counting by tens and ones instead of by finding where the farthest endpoint corresponded to a numeral on the measurement strip. For example, when measuring the length of a table, many students laid the measurement strip next to the edge of the table and counted each collection of 10 from the beginning of the table—"ten, twenty, thirty"—rather than simply read off the numeral 30. Thus, measuring with the measurement strip for many students appeared to have evolved out of their participation in prior mathematical practices in which they iterated single or collections of units.

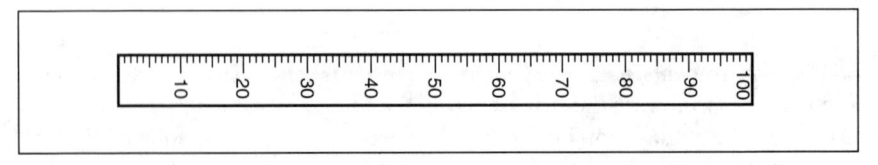

Figure 5.10. The measurement strip.

Very quickly, students abbreviated their activity of counting up on the measurement strip, and the fact became taken-as-shared that the length of an item could be measured by laying down the measurement strip alongside the item and simply reading off the numeral corresponding to the position of the farthest endpoint. Also taken-as-shared was that when students measured the length of an item and read off a number, that number signified the distance that extended from the beginning of the measurement strip to the line signifying the end of the item. Students now seemed to take-as-shared that they were acting in a spatial environment in which distance was already partitioned. Distance no longer had to be partitioned in an activity by physically iterating. Rather, the fact that laying down the measurement strip simply specified the measure of an already-partitioned spatial extent was taken-as-shared. Students seemed to be acting in an environment in which an item was already partitioned—that is, it already had a measure—and the measurement strip simply specified the measure. This taken-as-shared interpretation was significantly more sophisticated than in prior mathematical practices and constitutes operational measurement as defined by Piaget et al. (1960).

Meagan and Nancy. Both Nancy and Meagan came to interpret measuring in the way just described. Initially, however, Meagan laid down the measurement strip to find the length of a table and started by counting individual cubes, "one, two, three, ..." Nancy interrupted Meagan's counting and continued from where Meagan had left off, counting by fives. Meagan also counted by fives until they had "iterated" the length of the table. Hence, initially, for Meagan the measurement strip did not signify the result of iterating strips of 10. The very next day, however, she stopped counting by ones and, finally, no longer counted cubes at all. Rather, she simply located the appropriate place on the measurement strip and read off the measure of the item. This participation seemed to be supported by the physical record that was inherent in the measurement strip. The strip itself became a record of the units that Meagan had initially established or counted along the strip. Further, the numerals signified what she would have measured had she actually iterated a 10-strip. In reasoning in this manner, she contributed to the establishment of this fourth mathematical practice. As before, symbolizing and tool use were integral to her reasoning.

Other than in the foregoing example, Nancy never counted by tens or ones to establish the length of an item when using the measurement strip. The strip already signified, for her, the result of iterating a 10-strip, or smurf bar. When determining the measure of an item with the measurement strip, the measure signified the result of accumulating distances for both Nancy and Meagan. The length they had determined signified a distance that could be partitioned into tens and ones for them. Their reorganizations were made as they participated in the emerging mathematical practice of measuring with the measurement strip. Their reorganizations also constituted their contribution to the ongoing evolution of the fourth mathematical practice.

Conclusion of phase 4. In summary, the teacher built on the prior mathematical practice of iterating bars of 10 by asking students to draw paper strips 10 cans long. Then students engaged in measuring activities with these paper 10-strips. The teacher next taped the 10-strips together end to end to support students' construction of a measurement strip 100 cans long. A taken-as-shared way of measuring with the measurement strip was immediately constituted. Laying down the measurement strip seemed to partition the distance into cans, or cubes, that were themselves organized into composites of 10. That the spatial extension of an item was partitioned before the physical act of measuring took place was taken-as-shared, and when the measurement strip was laid down, it simply specified the measure. Measuring was now no longer tied to physically iterating to find an item's measure; rather, in a sense, the measurement strip signified the prior activity of iterating collections of 10. Measuring seemed to be routine at this point. Thus, the fourth mathematical practice involving measuring with a strip of 100 became established.

Phase 5—Reasoning with the Measurement Strip (5 Days)

Because the foregoing measuring activities became routine for students fairly quickly, the teacher posed problems that involved making comparisons of lengths on the measurement strip. At this point of the instructional sequence, the fifth mathematical practice, which involved reasoning with a strip of 100, emerged. The teacher introduced a scenario in which the wizard smurf was testing secret formulas that controlled the growth rates of sunflowers. The students were told that the wizard smurf was conducting an experiment in which he had created several secret formulas for speeding the growth of sunflowers. He noticed that seeds treated with various secret formulas grew into different-sized sunflowers. Students were asked to help the wizard smurf decide which formula would make his flowers grow taller. Wizard smurf needed to know how much taller each sunflower was than the others. One flower grew from a plain seed and was 51 cans tall. Seeds treated with secret formulas A, B, C, and D grew to be 45, 35, 61, and 70 cans tall, respectively. The students' task was to find out how much taller each of the flowers grown with secret formula were than the flower grown with a plain seed. This type of task differed from previous activities in that students no longer measured a physical item with the measurement strip. Previously, the strings and adding machine tape signified the spatial extent of the dragons, fish, wooden boards, and so forth, and students actually had to measure the objects. In the sunflower activity, the spatial extents of the various sunflowers were given numerically, and students had to specify the length on the measurement strip rather than measure it.

The emergence of mathematical practice 5—Reasoning with a strip of 100

Three days after the measurement strip had been introduced, the teacher posed the sunflowers problem. This instructional activity was the first one in which the students did not measure an item that was physically present. As a consequence, a new mathematically significant issue arose that continued to be the focus of discussion for the next two class periods. When the students specified the spatial extent of two sunflowers' heights, say, 51 and 45, some counted the spaces between the two numbers, whereas others counted the lines. The students who counted the spaces between the two specified heights obtained 6 as the difference (see Figure 5.11).

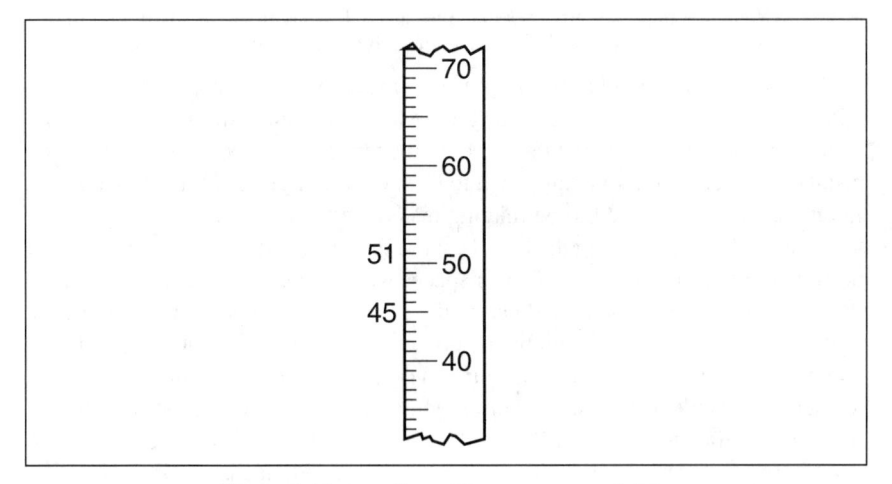

Figure 5.11. A portion of the measurement strip.

The students who counted the lines between the two heights started counting with the line beside 45 and stopped with the line beside 51, getting a result of 7. We thought that for students who were counting lines, their use of the measurement strip was divorced from their participation in prior mathematical practices. The gap between 45 and 51 on the strip did not signify for them a distance that could be covered by cubes. Instead, they simply counted the only things perceptually available—the lines. We therefore speculated that these students might be helped by engaging in a discussion in which their descriptions folded back to the context of measuring with cubes. In the excerpt that follows, the teacher asked students how much shorter the sunflower measuring 45 cans would be than the sunflower measuring 51 cans. A measurement strip was hung vertically at the front of the classroom, and Phil recorded where he thought the two sunflowers' heights would be on the strip (see Figure 5.11). Then he explained that the sunflower measuring 45 cans would be 7 shorter than the other would because he counted the lines between the two numerals.

Phil: From here down to here would be seven [*points to the line beside 51 and 45, respectively*]. Because one, two, three, four, five, six, seven [*counts the lines*].
Pat: I have a question. You're supposed to count spaces, not the lines, because it's ...
Phil: It would be the same because one, two, . . . six [*seems confused*]. I'm counting the lines because they're the same as spaces. Here's a line [*points to a line*], it goes down to here [*moves his finger down to the line below it*].

...

> Even if you count the lines, it's still like the space because you go down to here [*points to a line and moves his finger down to the line below it*].

...

Pat: Yeah, but there's more lines than one space. I'm saying there's two lines in one space because, see the front of one, the end of one. So there's two lines.

Phil and others seemed to be simply counting what was perceptually available without realizing they were counting an extra line. Our initial conjecture that students were not relating counting lines to the number of cans that partitioned the distance between the two heights had to be revised in light of Phil's explanation. In retrospect, Phil seemed to be relating his counting activity to counting spaces when he said, "I'm counting the lines because they're the same as spaces." In fact, he argued that counting either lines or spaces would give the same number. Thus, Phil and others seemed to be indicating the number of cans or the measure of the spatial extent between two heights by counting what was perceptually available without taking the extra line into account. In the preceding excerpt, the teacher tried to encourage students to justify their particular method in terms of what each line or space signified to them. For Pat, the lines signified the top and bottom of a cube and the space signified the distance covered by one whole cube. Thus, counting the lines gave an extra number because "there's two lines in one space." During this and other conversations, counting spaces to specify the measure of the spatial extension between two lengths became taken-as-shared. Students' explanations of their activity indicated that the idea had become taken-as-shared that a measure, such as 35, signified the distance starting from the bottom of the strip to the top of the 35th.

Meagan and Nancy. Meagan initially participated in this practice by first using a cube to mark the end of the spatial extensions of each of the two items being compared on the strip. Then she counted the number of cubes that fit exactly between the two cubes. For example, on the last day that the students used the measurement strip (day 31), the teacher asked them to measure one another's heights and record the results. The teacher, for her part, recorded on the board the results of each small group, for example, Chris 65 and Hilary 75; Lloyd 67 and Alex 72. Then the students were asked to work in their small group to find the differences between each pair of students' heights. Alice and Meagan used a measurement strip to find the difference between Chris's (65) and Hilary's (75) heights. Meagan placed a cube in the space signifying the 66th cube and a cube in the space signifying the 76th cube, as if she were marking the top of the two students' heights. Meagan then counted the number of spaces on the measurement strip between the two cubes, getting a total of 9 cubes (see Figure 5.12).

This example illustrates that at the very least, Meagan could specify the spatial extensions of two items on the measurement strip, that is, Hilary's and Chris's heights. These specified spatial extensions signified composites units, for example, the height stretching from 0 to 66. When it came to quantifying the difference (the

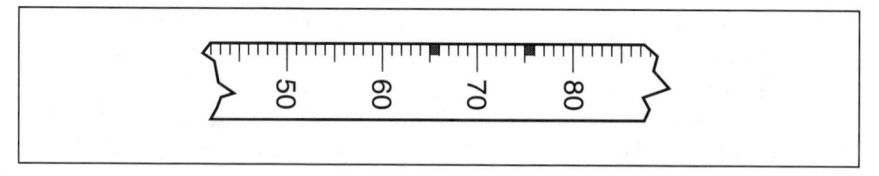

Figure 5.12. Two cubes marking the 66th and 76th spaces.

distance) between the two heights, however, she did not see Chris's height (65) as nested in, or as a part of, Hilary's height (75). Rather, Meagan seemed to be *measuring* the gap between the two cubes, and that gap did not appear to signify an entity in and of itself. In other words, she measured the gap between the two cubes instead of reasoning about the difference between two quantities. This approach indicates that Meagan did not seem to be reasoning in part-whole terms in this instance, that is, Chris's height was not nested in Hilary's height, with the remaining space viewed as an quantity to be measured.

Meagan's reasoning with symbols also contributed to the emergence of the fifth mathematical practice. In the subsequent whole-class discussion, the teacher asked Meagan to show how tall Lloyd (67) and Alex (72) were using a measurement strip that had been taped vertically at the front of the classroom. Meagan pointed to the line beside the numeral 67 and explained that Lloyd would be that tall, indicating the length from the floor to where she has marked 67. The teacher drew a short horizontal line at the line indicating Lloyd's height. Meagan indicated that Alex's height began from the floor to the line beside the numeral 72, and again the teacher drew a horizontal line to mark it. Note that the teacher's symbols served the same purpose as the cubes Meagan had used to mark Chris and Hilary's heights in the previous example. Then, to find the difference between Lloyd's and Alex's heights, Meagan placed a smurf bar between the two horizontal lines the teacher had drawn, broke off six cubes, and placed the six cubes so that they fit exactly between the two lines. The space between the two marks did, in fact, signify the difference between the two heights. This activity, however, is consistent with her activity in the prior episode, in that she did not reason with the two heights in part-whole terms. Rather, Meagan empirically measured to find the difference between the two marks.

Several students objected to the result and argued that the difference between the two heights was only 5 cans long. For example, Alex reasoned that 3 more cubes from 67 would be 70, and 2 more cubes made 5 cubes altogether. Hence, Alex took the gap between 67 and 72 as a measurable distance, as well, but did not actually measure to find the result. Other students simply counted 5 single cubes by ones. Phil argued, however, that because Meagan had measured the space and gotten 6, Alex must have counted incorrectly. The teacher then redrew the lines so they were straighter and more accurate, and when Meagan remeasured in the same manner as before, she found that exactly 5 cans fit. The mathematically significant issue that was the focus of this conversation centered on how to quantify the gap between the two heights—actually measuring versus reasoning. Meagan, for her part, drew on her prior participation in mathematical practice 4 by actually measuring the space with cubes to solve the task at hand. Alex, for his part, could take that gap as a measurable space and reason with it. Both Alex and Meagan could specify the two spatial extents very easily on the measurement strip by simply pointing to the lines corresponding to 67 and 72. The spatial extension that signified the physical length of an object seemed to be a partitioned distance for them; the distance signified by the measurement strip was already partitioned, and when Meagan measured with the strip, she was simply structuring that distance into composites of tens and ones. When attempting to reason with the spatial extensions that she had specified on the measurement strip, however, those spatial extensions did not appear to be nested in one another, that is, the two spatial extents were not related in part-whole terms and, therefore, she measured the space with cubes. In the conversation described in the foregoing, Meagan contributed to the constitution of a taken-as-shared interpretation that the measure of an item signified an objective property of that item—65 signified Lloyd's height and was now beyond justification.

Nancy, in contrast, participated in this practice by reasoning in part-whole terms. For example, on the 30th day of the instructional sequence, the teacher gave each pair of students a sheet of paper with pictures of various pieces of furniture and their respective measures, for example, box, 42 cans; television, 24 cans. The students were told that the smurfs wanted to stack the furniture items on top of one another so they could minimize the amount of storage space in a room that was 60 cans high. The following episode describes how Nancy typically reasoned about this problem situation. In one instance, Nancy decided to stack a box 42 cans high first. She identified the line beside the numeral 40 and counted up two more spaces, "forty-one, forty-two." She explained, "That's the forty-two cans. Here, all the way to here [*motions from 0 to 42*)]. It's the box." Clearly, the spatial extension that she had specified signified the height of the box. Then she counted by ones starting with the 43rd space to see how much room remained to stack another item on top of the box, "one, two, three, . . . , eighteen." After consulting the activity sheet, she explained, "We can put the lamp [*17 cans high*]. Here's the one more space that's left [*pointing to the 60th space*]." When a researcher asked her where the lamp was, Nancy replied, "from right here [*points to the line signifying the top of the 42nd can*] to here [*points to the line signifying the top of the 59th can*]." From Nancy's activity in this example, she appeared to be able to specify the spatial extension of two items on the measurement strip, and, in turn, reason with these extensions. The space from 42 to 59 signified, for her, the spatial extension of the lamp. Furthermore, the two spatial extensions together—the distance from 0 to 59 as well as each of the two spatial extensions themselves—were nested in the height of the storage room, 60. She reasoned that the box and lamp would fit inside the room with only one can of distance left over. In fact, when Meagan suggested stacking the television (24 cans high) on top of the box, Nancy argued that the room would not be high enough. She seemed to be able to imagine the spatial extension signifying the height of the box (42) being nested in the spatial extension signifying the height of the room (60) with 18 cans of space remaining. She therefore automatically eliminated furniture that measured over 18 cans high. This realization indicated that she was envisioning nesting two or more spatial extensions within the spatial extension signified by the height of the room and, in turn, could identify the amount of space left over after stacking the furniture. The important point is that Nancy viewed the unknown difference as a quantity whose measure she then found.

A form-function shift. Stacking problems such as those in the foregoing seemed to become relatively routine. Possibly, the idea that the measure of an item signified an objective property of that item had become taken-as-shared. In other words, a reversal between the spatial extent and its measure had taken place. Before, the spatial extent of an item was physically present and the goal was to find its measure by laying the measurement strip beside it. Now, the measure was given and the goal was to use the measurement strip to specify the spatial extent signified by its measure. This change constitutes what Saxe (1991) terms a *form-function shift*. In this way, the measure of an item signified an objective property of an item because the measure was not being found but was being used, that is, taken as a given, to specify an item's spatial extent. The spatial extent now signified the height of a stool or the height of a chair, and these heights took on a life of their own, so to speak— they could be used to compare and combine with other heights. As can be seen, the taken-as-shared way of reasoning with the measurement strip went beyond simply specifying already partitioned spatial extents on the strip. The practice that was constituted involved working from someone else's measure instead of actually finding the measure itself.

ANALYZING STUDENTS' LEARNING IN SOCIAL CONTEXT

The first issue discussed in the literature review was that of locating students' development of measuring conceptions in the social context of the classroom. Previous research (e.g., Clements, Battista, & Sarama, 1998; Inhelder, Sinclair, & Bovet, 1974; Piaget et al., 1960; Smedslund, 1963) typically analyzed students' solutions to interview tasks from primarily an individualistic viewpoint. Thus, the main goal of this chapter was to analyze the first graders' development of measuring conceptions as it occurred in social context. To this end, we viewed learning as an act of participation in the local mathematical practices of the classroom community. We detailed the emergence of five classroom mathematical practices that served to document the immediate social context of the individual students' learning. We presented two case studies in conjunction with the documentation of the collective learning. The purpose of presenting the two simultaneously was to show that (1) students' learning occurred as they participated in these emerging practices, and (2) the mathematical practices emerged as students, often the target students, contributed to them. A traditional cognitive analysis would have treated the target students' learning independently of the local classroom community by describing their development in terms of cognitive reorganizations with little attention to the social context, that is, to the collective argumentation. Although we described the cognitive development of each target student by drawing on the cognitive constructs of Piaget and his colleagues and of Steffe et al. (1988)—for example, accumulation of distances and partitioning—we cast each instance of learning as an act of participation in the mathematical practice that was either emerging or was established. For instance, Meagan made certain reorganizations as she participated in the whole-class discussion in which certain practices were established. Furthermore, the mathematical practices were documented by analyzing the way in which structuring distance evolved over time and became normative in the classroom discourse.

When we talk about the mathematical activity of a community, we refer to the taken-as-shared activity of the community rather than the overlap of the meanings of all its individual members (Voigt, 1996). An analysis that focuses on the overlap in individual meanings would constitute a cognitive analysis of the development of a *plurality* of individuals rather than the mathematical activity of a *single* community. Thus, when describing the classroom mathematical practices, we made claims about the mathematical practices of a community but not about how any one child was reasoning. Instead, we described ways of reasoning that became normative over time as indicated by the taken-as-shared mathematical activity observed in the public discourse. The documentation of the classroom mathematical practices reflects the way in which taken-as-shared ways of reasoning, arguing, symbolizing, and using tools evolved as the measurement sequence was enacted. When we focused on individual students' participation in these mathematical practices, we made claims about each student's personal way of reasoning, but always in the context of the taken-as-shared practices of the community.

One final issue that should be addressed here relates to the usefulness of the classroom mathematical practice as a construct for documenting the learning of a community. As previously mentioned, Cobb and Yackel (1996) first developed the construct of the classroom mathematical practice as a way to account for the mathematical learning of a classroom community over long periods of time. Thus that construct is useful for describing the emerging social situation of the mathematical learning that is jointly constituted by the teacher and the students. The analysis we have presented should clearly show that the classroom mathematical practices are not to be viewed as emerging independently of the previously established practices. In other words, each practice grew out of practices previously established by the classroom community. Also, the mathematical practices did not suddenly appear fully formed. The evolution of the taken-as-shared meanings and interpretations involved an analysis and discussion of each of the mathematical practices with the intent of documenting the process of their emergence. Thus the analysis of the classroom mathematical practices served not only as documentation of the evolving social situation in which the students were participating but also, retrospectively, as a summary of the mathematical content that emerged over the course of the measurement sequence. Furthermore, the documentation of the collective mathematical practices, the evolving tool use, and the quality of whole-class discourse provide the necessary backdrop for teachers who wish to incorporate aspects of the instructional sequence in their own socially situated classrooms.

In contrast with a purely cognitive analysis, some social theories might have involved analyzing the learning of the classroom independently of the variety of students' thought processes. In our view, the collective mathematical practices emerged only as students contributed their personal mathematical interpretations. In sociocultural traditions, the teacher guides students to fuller participation in broader practices of the mathematical community. In this sense, mathematical practices are fixed a priori and the teacher's goal is to guide students' enculturation into these practices. In our use of the term, *mathematical practices* are not fixed prior to experimentation but emerge as students contribute to them. Thus, classroom mathematical practices are normative for the particular classroom social structure in which they are established and draw on the diversity of students' contributions. The role of the teacher in this type of classroom is to guide students' reasoning so that their practices become more consistent with those of wider society.

CONCLUSION

In this chapter, we have used a sample analysis from a first-grade classroom to discuss the theoretical nature of coordinating social and individual aspects of learning. We presented the evolution of the collective mathematical practices and complementary case studies of two students' individual development. We argued that although five practices emerged in the collective discourse, individual students' development constituted acts of reasoning that contributed to the evolution of these practices. Toulmin's (1969) model of argumentation was used as a methodological tool to determine, from the observers' standpoint, when practices appeared to be established. Finally, we reflected on one of the overarching themes of the monograph—analyzing students' learning in social context—by contrasting this type of analysis with both traditional cognitive and sociocultural analyses. During that discussion, we argued that although traditional analyses are useful, neither traditional nor sociocultural analyses alone gives a full explanation of the process of learning.

Although the main focus of discussion has been on the theoretical implications of learning in social context, we have argued the practical merits of such an analysis elsewhere (Cobb, chapter 1 of this monograph; Gravemeijer et al., chapter 4 of this monograph). The practices and case studies provide a background against which to make revisions to the measurement sequence and to incorporate the instructional sequence within one's own classroom. We discuss these characteristics of mathematical practices further in the next chapter.

REFERENCES

Bauersfeld, H. (1988). Interaction, construction, and knowledge: Alternative perspectives for mathematics education. In T. Cooney & D. Grouws (Eds.), *Effective mathematics teaching* (pp. 27–46). Reston, VA: National Council of Teachers of Mathematics, and Hillsdale, NJ: Erlbaum.

Clements, D., Battista, M., & Sarama, J. (1998). Development of geometric and measurement ideas. In R. Lehrer & D. Chazan (Eds.), *Designing learning environments for developing understanding of geometry and space* (pp. 201–226). Hillsdale, NJ: Erlbaum.

Cobb, P. (1994). Where is the mind? Constructivist and socioculturalist perspectives on mathematical development. *Educational Researcher, 23*(7), 13–20.

Cobb, P., & Yackel, E. (1996). Constructivist, emergent, and sociocultural perspectives in the context of developmental research. *Educational Psychologist, 31*, 175–190.

Inhelder, B., Sinclair, H., & Bovet, M. (1974). *Learning and the development of cognition.* Cambridge, MA: Harvard University Press.

Piaget, J., Inhelder, B., & Szeminska, A. (1960). *The child's conception of geometry.* New York: Basic Books.

Saxe, G. (1991). *Culture and cognitive development: Studies in mathematical understanding.* Hillsdale, NJ: Erlbaum.

Smedslund, J. (1963). Development of concrete transitivity of length in children. *Child Development, 34*, 389–405.

Steffe, L. P., Cobb, P., & von Glasersfeld, E. (1988). *Construction of arithmetical meanings and strategies.* New York: Springer-Verlag.

Toulmin, S. (1969.) *The uses of argument.* Cambridge, England: Cambridge University Press.

Voigt, J. (1996). Negotiation of mathematical meaning in classroom processes. In P. Nesher, L. P. Steffe, P. Cobb, G. Goldin, & B. Greer (Eds.), *Theories of mathematical learning* (pp. 21–50). Hillsdale, NJ: Erlbaum.

Chapter 6

Continuing the Design Research Cycle: A Revised Measurement and Arithmetic Sequence

Koeno Gravemeijer
Freudenthal Institute
Janet Bowers
San Diego State University
Michelle Stephan
Purdue University Calumet

INITIATING A NEW DESIGN RESEARCH CYCLE

In chapter 5, we analyzed the development of the first graders' conceptions of measurement and, in so doing, completed the second phase of the Design Research Cycle. In this chapter, we initiate a new cycle by reflecting on the revised sequence. This discussion focuses on the three themes that were discussed during the literature review. The first theme involved analyzing students' development of measuring conceptions as it occurs in social context. The second theme concerned the need to delineate the role of tools in supporting students' understanding. We address this issue by documenting the evolution of a chain of signification (Walkerdine, 1988), which is based on our analyses of the taken-as-shared meanings that emerged as students used a variety of tools in the measurement sequence. The third theme centered on proactively supporting students' development through integrating research and instructional design. We address this issue by reflecting on our prior work to develop suggested revisions for ensuing phases of Design Research.

Theme 1: Analyzing Students' Learning in Social Context

In the previous chapter, we argued that identifying the collective mathematical practices and coordinating them with individual case studies give a fuller account of the learning of the first-grade community than traditional analyses that cast learning as primarily an individualistic endeavor. The main emphasis of this chapter is to demonstrate that the mathematical practice as a theoretical construct has usefulness as it relates to instructional design, as well. As Gravemeijer et al.

(chapter 4 of this monograph) noted, mathematics research has seen a shift in instructional research interests from merely stating that one instructional approach is better than another on the basis of some statistical measures. Our concern with these types of comparative studies is that they provide little explanation as to how the instruction was realized in social context, and, therefore, teachers would have difficulty knowing why the approach did not work in their particular situation. Furthermore, because classrooms are socially diverse entities, instructional approaches that offer recipes for implementation cannot do so for every possible encounter. In contrast, we suggest that the products of instructional design research should present rich descriptions of the instructional tasks, classroom activity structure, tools and discourse that can lead to the development of desirable mathematical practices. As Cobb and his colleagues have noted elsewhere (Cobb, Stephan, McClain, & Gravemeijer, 2001), the pragmatic power of the classroom mathematical practice lies in the fact that being able to anticipate the learning route of each of 25 individual students is not feasible for a teacher. One can, however, think of guiding the discourse and tool use of a class in a way that allows for, and capitalizes on, the diversity of students' thinking. Therefore, the hypothetical learning trajectory in chapter 4 and the resultant instructional theory outlined in this chapter can be thought of in collective terms. The learning route is that of a class, not of individual students. Therefore, we present the resulting instructional theory on measuring and arithmetic in this chapter, noting that these theories were designed using the mathematical practices as the backdrop. In other words, the designer uses the analysis in chapter 5 detailing the mathematical ideas and discussion topics that became taken-as-shared measurement practices to formulate the rationale and intent of the resulting instructional theory found in this chapter. To summarize, the power of the classroom mathematical practice from the standpoint of the designer is that the practices serve as the basis for outlining the potential mathematical ideas and conversation topics during the enactment of the sequence. The usefulness from a teacher's standpoint is that these potential practices can be used to lead a community, rather than individual students, to sophisticated mathematical reasoning.

Although we describe the sequence in terms of the collective mathematical reasoning that can emerge as teachers and students engage in these instructional activities, we do not neglect individual students' reasoning. On the contrary, the case studies presented a rich account of how two students participated in, and contributed to, the emerging practices. Thus, any teacher implementing the instructional theory discussed in this monograph can use the theory to anticipate potential collective shifts in mathematical discourse that are made possible by students' diverse ways of reasoning and of contributing to the emerging taken-as-shared practices. The case studies provide examples of students' different ways of reasoning that a teacher might expect from her or his own students as the sequence is implemented in the classroom. The teacher can use these examples to anticipate the contributions from students that she or he wants to draw on in classroom discussions.

Theme 2: Documenting the Role of Tools with a Chain of Signification

A second, but related, theme from the literature review was that of documenting the interplay between the development of meaning and that of symbolizing and tool use. The rationale for this aspect of the analysis reflected the growing realization in the mathematics education research community that learning does not occur apart from reasoning with symbols and tools. Further, much of the prior literature on students' development of measuring conceptions attributed a limited role to reasoning with measurement tools. Given that measuring inherently involves using both conventional and nonconventional measurement devices, this study offers an opportunity to investigate the role of tool and symbol use in students' mathematical development. Along with Meira (1998), we use the word *tool* to mean any physical device, such as a ruler, computer microworld, or a calculator, and the word *symbol* to refer to semiotic systems, such as graphs, tables, icons, or drawings.

A review of the emergence of the classroom mathematical practices reveals that the use of tools and symbols was integral to the mathematical activity of the community. As figure 6.1 illustrates, the emergence of the first four mathematical practices occurred as students acted with different tools and symbols (see Figure 6.1). As indicated by the transition from the fourth to the fifth mathematical practice, however, a new mathematical practice can emerge as students reason with the same tool. Recall from the analysis that the transition from the fourth to the fifth practice involved shifting from measuring with the measurement strip to reasoning with specified quantities—spatial extensions—on the measurement strip. Thus, as the taken-as-shared function of the measurement strip changed, a new mathematical practice arose. The analysis makes clear that the communal learning, the emergence of mathematical practices, did not occur apart from reasoning with tools. In the previous analysis of classroom mathematical practices, however, the focus was on coordinating individual and social perspectives to give a full account of learning, with explicit focus on the role of tools in this process fading into the background. In this section, we bring the evolution of tool use to the fore by elaborating the taken-

---

CLASSROOM MATHEMATICAL PRACTICES

1. Measuring by covering distance
2. Partitioning distance with a collection of units
3. Measuring by accumulating distance
4. Measuring with a strip of 100
5. Reasoning with a strip of 100

Figure 6.1. The evolution of the classroom mathematical practices.

as-shared meaning that developed as students reasoned with a variety of tools. To do so, we detail both the meaning and the taken-as-shared goals and purposes that evolved. Such an analysis is crucial for documenting the taken-as-shared imagery and activity with tools that are part of the measurement instructional theory presented at the end of this chapter.

In this teaching experiment, the general notion of a ruler—that is, the general form and structure of a ruler although without the inch markings of a conventional ruler—served as the overarching model (Gravemeijer, 1998). As described in chapter 4, a model can take many forms throughout an instructional sequence. One way to describe the various manifestations of a model is to describe the evolving taken-as-shared meanings associated with the use of the tools and symbols. Cobb, Gravemeijer, Yackel, McClain & Whitenack (1997) and Gravemeijer (1998) both suggest documenting this evolutionary process by describing the emergence of a *chain of signification*. A chain of signification, as described by Walkerdine (1988), captures the process by which taken-as-shared activity with symbols and tools comes to signify the use of another. When one symbol or tool comes to signify a previous symbol or tool, Walkerdine states that one serves as a signifier for the other. This relationship forms a signified-signifier pair and constitutes a link in a more global chain of signification. For example, in the measurement sequence, students initially paced the length of items and subsequently made records of this activity, for example, when the teacher placed masking tape to mark each pace. Thus, the first signified-signifier combination that became constituted as the first mathematical practice emerged involved pacing and records of pacing (see Figure 6.2).

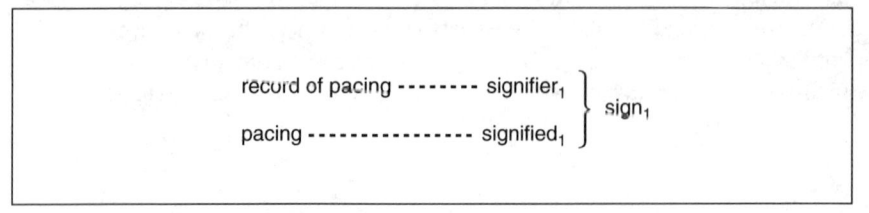

record of pacing -------- signifier$_1$ ⎫
⎬ sign$_1$
pacing ---------------- signified$_1$ ⎭

Figure 6.2. The first link in a chain of signification.

More specifically, distance, as structured by records of paces, served as the signifier for pacing. As we learned from the analysis of the classroom mathematical practices, distance was initially something to be filled by paces and then by records of paces. Thus, as the first mathematical practice emerged, records of paces came to signify pacing activity.

Figure 6.2 illustrates that the first link in the chain of signification is that between records of pacing and pacing, designated as {records of pacing/pacing}. As Cobb et al. (1997) note, one advantage of identifying a chain of signification is that a signifier-signified combination can, in turn, be signified by other symbols as the interests of the community change. Specifically, in the instance of the measurement sequence, the constitution of the first link in the foregoing (Figure 6.2) developed as the interest of the class involved deciding how one could measure with one's feet in accurate and spatially legitimate ways. As students measured with the footstrip, the taken-as-shared interests of the community changed and a second mathematical practice emerged that involved partitioning distance with a *collection* of units. No longer were students concerned only with measuring and counting single units; the taken-as-shared purpose of the community changed to partitioning distance with collections of units—in the footstrip tool—that could not be physically separated. As this practice was constituted, a second link in the chain of signification was established and can be illustrated as shown in Figure 6.3. In this case, the footstrip served as a signifier—signifier 2—for the first signified-signifier pair. Walkerdine (1988) uses the word *sign* to refer to a signifier-signified pair. Thus, {record of pacing/pacing} constitutes the first sign, which is signified by the taken-as-shared activity and purposes for measuring with the footstrip.

footstrip ---------------------------- signifier$_2$

record of pacing ---- signifier$_1$

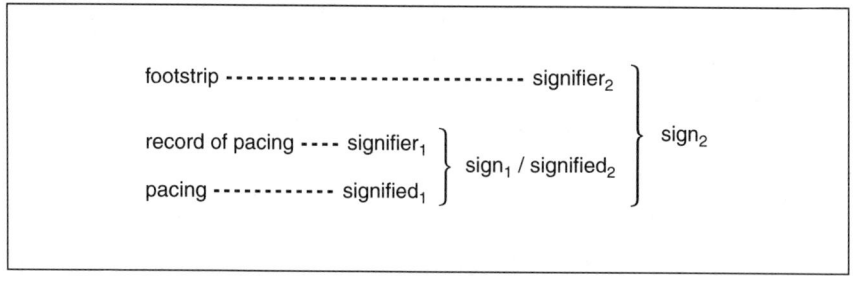

pacing ----------- signified$_1$

sign$_1$ / signified$_2$

sign$_2$

Figure 6.3. A second link in the chain of signification.

As the first and second mathematical practices emerged, the chain of signification exemplified in Figure 6.3 was constituted. These links were constituted as students discussed their interpretations of the result of measuring. As this link was constituted, the relationship between the signifier and the signified was not one in which one symbol merely replaced the other. Rather, the mathematical practices became established as students reasoned with subsequent tools and as the community's interests changed; this outcome, in turn, changed the nature of the activity with previous signifieds; thus, the nature of distance evolved for students. In this way, signified entities are said to slide under succeeding signifiers rather than be masked or replaced by them (Cobb et al., 1997; Walkerdine, 1988). As students measured with the footstrip, mentally structuring distance into composites of 5 became taken-as-shared. Measuring became divorced from activity, that is, the idea was taken-as-shared that a numeral, such as 25 paces, signified the measure of a spatial extension apart from physically measuring to create the measure. Therefore, as students measured with the footstrip, not only did the idea become taken-as-shared that the footstrip signified the previous sign, but also the nature of distance again evolved. This example illustrates the interplay between the taken-as-shared mathematical activity of the classroom community and their evolving taken-as-shared symbolic meanings. In other words, as the mathematical practices evolved and the interests of the community changed, new links in the chain of signification emerged.

A second chain of signification began to emerge when the teacher introduced the smurf scenario. We say that a new chain began because the background scenario changed from measuring with feet and collections of feet to measuring with single Unifix cubes for the purposes of helping smurfs perform their measuring tasks. Also, the initial tasks in the new scenario recapitulated the first chain; the new scenario began with iterating single units and moved to iterating collections of units. At the beginning of this scenario, students used Unifix cubes as substitutes for food cans. The understanding that a measure, say, 27 cans, signified the length of an item became taken-as-shared fairly quickly. Thus, the first link in the second chain consisted of food cans signified by Unifix cubes (see Figure 6.4).

Unifix cubes -------------- signifier

food cans ---------------- signified

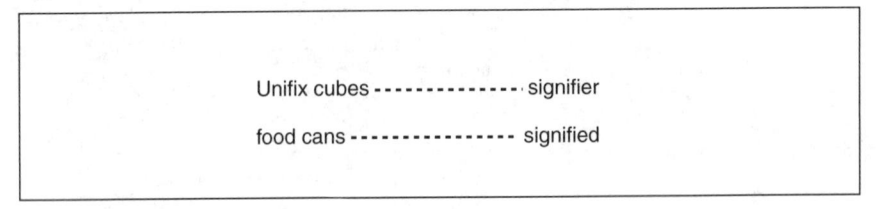

Figure 6.4. The beginning of a second chain of signification.

The fact that this link was constituted fairly quickly can be explained by taking note of the students' participation in prior mathematical practices during which the first chain of signification was constituted. Recall that we had attempted to introduce the smurfs scenario earlier, prior to the students' construction of the footstrip. When we did so, it became apparent that most students had difficulty taking a measure, such as 41 cans, as a given for the height of a wall in the smurf village. We conjectured that the inability of the class to take the result of measuring—41 cans—as a given for the length of an item was taken-as-shared; rather, measuring was grounded in the activity of iterating, that is, iterating to create the measure 41 cans. As the students engaged in activities where they iterated the footstrip and participated in discussions where interpreting the result of measuring was an explicit topic of conversation, the second mathematical practice emerged. As discussed previously, distance was no longer something to be covered with an exact number of footstrips. Rather, as the second mathematical practice was established, the fact that the results of measuring could be divorced from the activity of counting became taken-as-shared. *Twenty-five* signified the number of times that one would pace to measure the length without actually carrying out the action. In other words, 25 specified an amount of distance that could be measured, or a potentially measurable distance. Therefore, as the link between the footstrip and its signified became established, the nature of distance evolved such that taking a measure, such as 41 cans, as a given for the length of an object was now taken-as-shared. The first time we introduced the smurf scenario, only the first link of the chain (see Figure 6.2) had been established. At that point, the idea was taken-as-shared that distance signified something to be covered in activity, something to be created by physically measuring. For that reason, the introduction of the smurf scenario was inappropriate at that point, and the students engaged in activities with the footstrip. In that way, the constitution of the first chain of signification was integral to the emergence of the second chain (see Figure 6.5).

As the instructional activities changed from measuring with individual cubes to iterating a smurf bar, measuring by accumulating distances emerged as the third mathematical practice. Further, issues that emerged as topics of conversation now focused on the students' interpretation of distance rather than their method of measuring per se. For example, the students were asked to describe how they interpreted a measure, such as 20. Many students explained that 20 was the distance accumulated by iterating a bar twice and saying, "ten and ten is twenty." Thus, the interests and activity of the community had changed from finding methods of

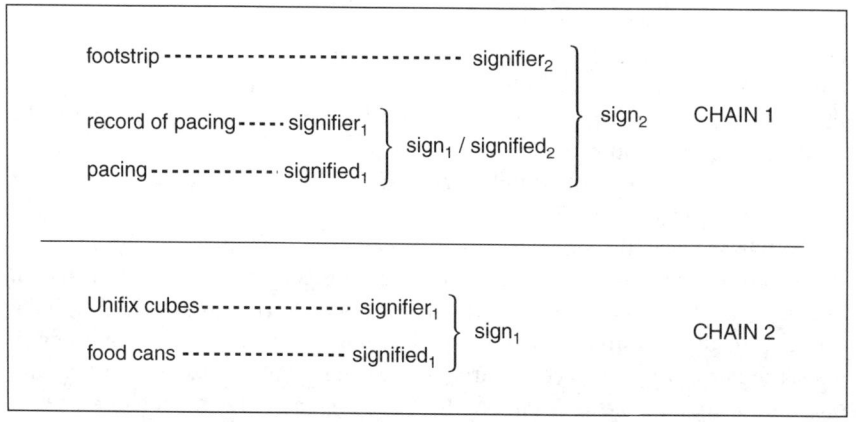

```
footstrip ---------------------------- signifier₂  ⎫
                                                   ⎬  sign₂     CHAIN 1
record of pacing ----- signifier₁ ⎫                ⎪
                                  ⎬ sign₁ / signified₂
pacing ------------- signified₁   ⎭


Unifix cubes -------------- signifier₁ ⎫
                                       ⎬ sign₁              CHAIN 2
food cans ---------------- signified₁  ⎭
```

Figure 6.5. The emergence of two interrelated chains of signification.

iterating and counting to solve the measuring dilemmas of the smurfs to interpreting and organizing distance in measures of tens and ones. As this third mathematical practice was established and the interest of the class changed to describing particular ways of organizing distance mathematically, a further link in the chain of signification was formed. Now the food can–Unifix cube combination, designated as {food cans/Unifix cubes}, became signified by the smurf bar (see Figure 6.6).

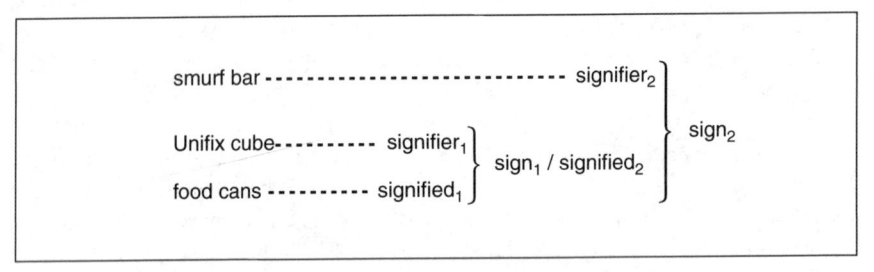

$$\left. \begin{array}{l} \text{smurf bar} \dashrightarrow \text{signifier}_2 \\ \left. \begin{array}{l} \text{Unifix cube} \dashrightarrow \text{signifier}_1 \\ \text{food cans} \dashrightarrow \text{signified}_1 \end{array} \right\} \text{sign}_1 \,/\, \text{signified}_2 \end{array} \right\} \text{sign}_2$$

Figure 6.6. A second link in the second chain of signification.

The first chain of signification, involving the footstrip and pacing, was also paramount to both the food can–Unifix cube link and the constitution of the link between the smurf bar and the food cans or Unifix cubes. As students measured with the footstrip, distance was structured in collections of 5 and single paces. As a result of discussing their engagement in these activities, measuring became divorced from pacing activity and this way of reasoning about and structuring distance became taken-as-shared in the classroom community. Because of this historical construction of an understanding of distance, measuring with single Unifix cubes became taken-as-shared almost immediately. Further, iterating a collection of 10 cans was, in some sense, similar to measuring with a collection of 5 paces. Although the number in the collection was different—10 instead of 5—the underlying image of distance as divorced from physically measuring was the same in the context of the smurfs. Thus, the first chain of signification was paramount to the emergence of both the first and second links of the second chain of signification (see Figure 6.7).

The link between the smurf bar and the food cans or Unifix cubes was constituted as the third mathematical practice, measuring by accumulating distances, emerged. The emergence of the third mathematical practice signals a shift in the interests of the classroom community. The interest had changed from simply discussing how to count paces to structuring distance into collections of tens and ones, for example, 20 is two tens. At the beginning of the instructional sequence, the interest of the community was on how to physically measure an item. Now the

```
┌─────────────────────────────────────────────────────┐
│                    Chain 1                          │
│                   ╱      ╲                          │
│   {food cans ---------- Unifix cubes} ---------- smurf bar    (Chain 2) │
└─────────────────────────────────────────────────────┘
```

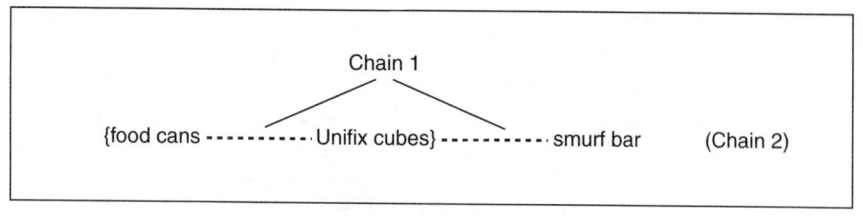

Figure 6.7. An emerging chain of signification.

interests and conversations focused on organizing and structuring distance mathematically. Thus, the meaning of the previous sign combinations evolved once the second chain was established. Further, the nature of distance evolved, in that now the result of measuring signified an accumulation of distance.

As the fourth practice of measuring with a strip of 100 emerged, two new signifying links were formed. First, in the fourth practice, the recognition became taken-as-shared that measuring with the 10-strip served as a symbolic record of iterating a smurf bar. Recall that this meaning emerged as students discussed the utility and meaning of marking each successive food can on their strips. The interests of the students again had changed from measuring by iterating a bar to finding an efficient and useful way of marking the 10-strip so that it could be used as a measuring device without the presence of *any* Unifix cubes. Then the purpose of their measuring and discussions was to find ways of both structuring distance with the 10-strip and coordinating measuring with tens and ones by accumulating distances. Although their practical goals were to find the measure of objects in the smurf village, their mathematical goals had changed to interpreting and refining their methods of counting their iterations of the 10-strip. A link was thus formed in which measuring with a 10-strip came to signify the meanings associated with iterating a smurf bar (Figure 6.8).

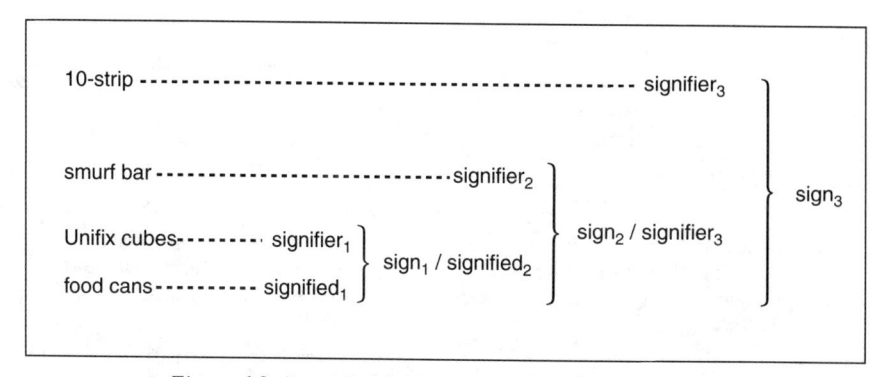

```
┌─────────────────────────────────────────────────────────────┐
│                                                             │
│   10-strip ------------------------------- signifier₃  ⎫    │
│                                                        ⎪    │
│                                                        ⎪    │
│   smurf bar ---------------------- signifier₂  ⎫       ⎬ sign₃│
│                                                ⎪       ⎪    │
│   Unifix cubes -------- signifier₁  ⎫          ⎬ sign₂ / signifier₃ ⎪│
│                                     ⎬ sign₁ / signified₂            ⎪│
│   food cans --------- signified₁    ⎭                  ⎭    │
│                                                             │
└─────────────────────────────────────────────────────────────┘
```

Figure 6.8. A new link in the second chain of signification.

Measuring with the 10-strip soon gave way to the more efficient 100-strip and came to be signified by the results of measuring with the 100-strip. This new link again involved a shift in the purposes of measuring for the community. Instead of finding methods of iterating a tool to measure, now the agreement became taken-as-shared that the purpose of their measuring was to develop a way to interpret measuring with a strip that they did not need, for the most part, to iterate. Recall from chapter 4 that students often looked back (Pirie & Kieren, 1994) to measuring with smurf bars to interpret measuring with the 100-strip. Thus, the history of evolving links is strong, in that new links may be formed by drawing on mathematical practices formed in several previous links, not just the one immediately preceding it. As the fourth practice was established, the understanding became taken-as-shared that distance signified a partitionable unit that could be structured in collections of tens and ones. Further, the idea was taken-as-shared that distance had the characteristic or property of a measure that was independent of physically iteration. Laying down the measurement strip simply specified the item's measure. Thus, the global chain of signification that emerged over the course of the measurement sequence can be pictured in its entirety as in Figure 6.9.

The transition from the fourth to the fifth mathematical practice involved a shift from measuring with the measurement strip to reasoning with the measurement strip. This transition in practices and interests signaled a shift in the taken-as-shared understanding of distance. Now students specified spatial extensions of items on the strip and reasoned about the differences between lengths and heights. Following Saxe (1991), this transition can be seen to involve a form-function shift, in that now the function of the measurement strip has changed from measuring to reasoning. This form-function shift is also consistent with the "model of–model for" transition described by Gravemeijer (1994).

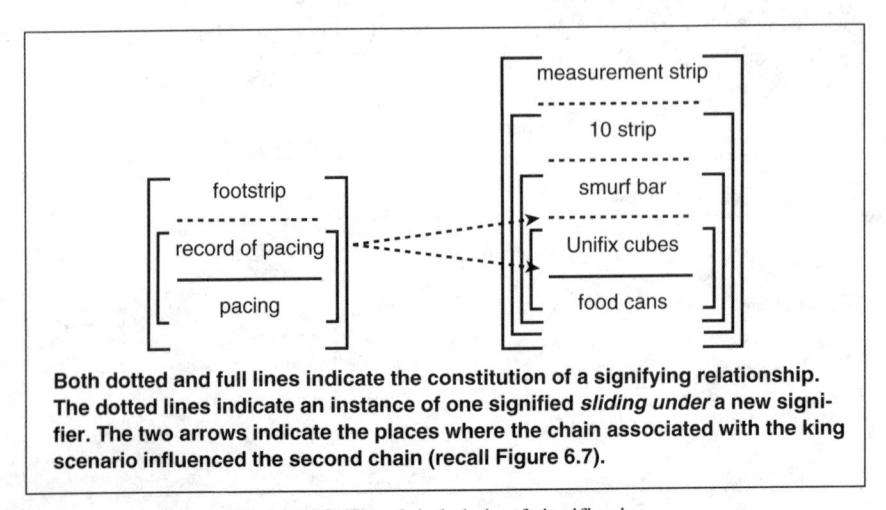

Both dotted and full lines indicate the constitution of a signifying relationship. The dotted lines indicate an instance of one signified *sliding under* a new signifier. The two arrows indicate the places where the chain associated with the king scenario influenced the second chain (recall Figure 6.7).

Figure 6.9. The global chain of signification.

Again, an important point to note is that the shift from measuring to reasoning with the measurement strip coincided with a change in the interest of the classroom community. The taken-as-shared goals shifted from measuring an item to finding ways to interpret and use the measurement strip to make comparisons among spatial extensions. This outcome is indicative of a general pattern: As the interests of the community change, that is, become more mathematical, new links in the chain of signification evolve. The realization is imperative that Figure 6.9 incorporates not only the change in physical tools but also the evolution of classroom mathematical practices and taken-as-shared interests and purposes that accompany new symbolizations.

Although the focus of the analysis in chapter 5 ended with students' construction of, and reasoning with, the measurement strip, two more tools were created and used by the students: the measurement stick and the empty number line. We have not included these two tools in the global chain of signification because we have not conducted formal analyses of the mathematical practices and taken-as-shared goals associated with using these two tools. We have, however, conducted an informal analysis to anticipate how or whether we can include these tools within a hypothetical learning trajectory. We discuss the importance of these two tools in two subsequent sections of this chapter: "Generalization" and "Revisions to the Sequence."

A semiotic analysis of this type brings to light the role that tools play in supporting students' learning along with the mathematical practices that are constituted during the evolving semiosis. Furthermore, the analysis emphasizes the evolving purposes and goals of the community that accompany the transitions involving new tools. Such an analysis brings to the fore the taken-as-shared symbolizations and meanings that accompany the mathematical practices and the taken-as-shared interests that evolve as the practices are constituted, both important aspects of classroom mathematical practices (Cobb, 1999). An additional benefit of the chain-of-signification analysis is that the analysis can shed light on why certain tools can be included or omitted in an instructional sequence. For instance, when we designed the initial measurement sequence, we did not intend to have students reason with a footstrip. The chain of signification indicated, however, that the taken-as-shared learning that emerged as students acted with the footstrip was paramount in the sequence of the use of two tools, the 10-strip and the measurement strip, that is, the 100-strip. Thus, the measurement sequence has now been modified to include the footstrip and other revisions as described in a subsequent section of this chapter.

Theme 3: Design Research

A third issue from the literature review concerns the relation between research and instructional design. Chapter 2 indicated that a number of the prior measurement studies used training methods to support students' development of measuring conceptions. Typically, researchers who engaged in training studies drew on

Piaget's cognitive theory to derive training tasks. Instruction on these tasks was then implemented, and the students were assessed to determine whether the training instruction had been effective (see Figure 6.10).

```
(Piaget's developmental) theory ⟶ training procedure ⟶ measured outcomes
```

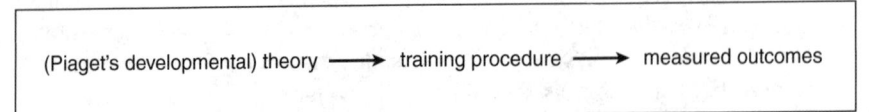

Figure 6.10. Training procedures.

The linearity in Figure 6.10 is based on a one-way process in which experimenters list a set of hypotheses prior to the study and then test their conjectures with pretests and posttests. The account of students' learning given in training studies provides evidence only as to whether the hypotheses were proved. In contrast, although developers who are engaged in the Design Research paradigm also make conjectures that are tested and revised, their intent in research is to construct the means of supporting development rather than to determine whether a conjecture was proved true or false.

The iterative cycles of Design Research can be viewed at two levels. At a global level, the first phase of a cycle begins when designers anticipate a global learning trajectory—rather than specific instructional activities—on the basis of results from previous research. Then the designers carry out an initial thought experiment in which they envision what students might learn as they participate in instructional activities, the quality of discourse and mathematical practices that supports learning, and the evolving cascade of inscriptions (Latour, 1990) or potential chain of signification (Walkerdine, 1988). Gravemeijer (1999) notes that in conducting this thought experiment, the developers formulate conjectures both about students' mathematical development and about the means of supporting it. In the second phase, the research team carries out a classroom teaching experiment in which the initial conjectures concerning the realization of the proposed sequence are tested and revised in action. Once the classroom teaching experiment concludes, a retrospective analysis, such as the one described in chapter 5, is conducted to inform the designers of any unanticipated interpretations that arose in the social setting of the classroom.

One interesting aspect of Design Research is that it is inherently recursive in nature. At a microlevel, daily minicycles are enacted during the research phase. For example, as the measurement teaching experiment took place, researchers used informal analyses of students' mathematical reasoning to make decisions about daily revisions of the instructional activities. Consequently, the resulting instructional sequence was shaped by our daily analyses of students' mathematical activity, which were made against the backdrop of a conjectured learning trajectory. Further, revisions to the sequence shaped students' subsequent mathematical activity. Therefore, we can now augment the initial Design Research figure introduced in chapter 2 by adding the minicycles shown in Figure 6.11.

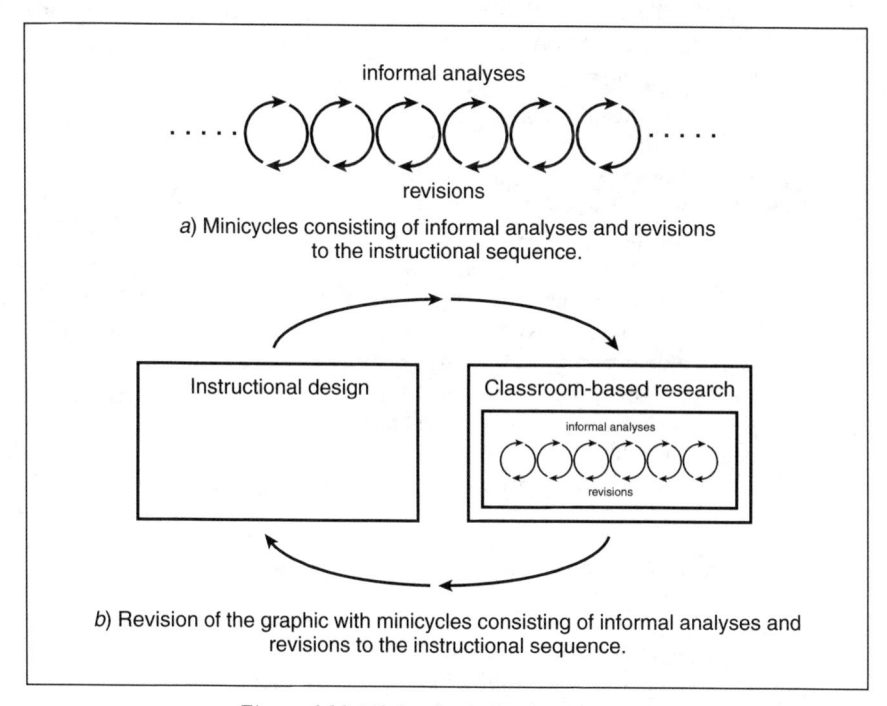

Figure 6.11. Minicycles in Design Research.

In summary, the documentation of the mathematical practices that became taken-as-shared in the community provided an account of the instructional sequence that emerged as a consequence of these daily modifications. As we documented the emergence of the mathematical practices, we took care to describe our role in the enactment of the instructional sequence. Where appropriate, we provided information concerning our conjectures about students' understanding and our subsequent decisions concerning further instructional activities to clarify our role in supporting students' learning. The analysis of the mathematical practices gave a situated account of the learning that was tied to students' participation in practices as they reasoned with tools. Thus the account of students' learning presented in this monograph was integrally tied to the means of supporting that learning. These analyses have led us to initiate a new cycle of Design Research by describing our revised sequence.

REVISIONS TO THE SEQUENCE

In this section, we first describe three revisions to the sequence and their accompanying rationales. We then present a table that incorporates our instructional design theory by listing the imagery, taken-as-shared interests of the community, and potential topics for discourse for each tool in the revised sequence. Our goal in constructing this table is to develop a heuristic for others attempting to envision a hypothetical learning trajectory.

Revision 1: The Introduction of the Footstrip

The genesis for this revision occurred the first time we introduced the smurfs scenario and asked students to show the height of a wall with Unifix cubes. Our intention was to use pacing activities only as a setting to introduce the idea of measuring. We conjectured that students would move easily from pacing to activities with the Unifix cubes, but, as documented in the analysis, they did not. We found that taking the result of measuring with the food cans as a given was not taken-as-shared, that is, the quantity of 41 cans did not signify the result of measuring but rather was created by physically pacing. As a consequence, we conjectured that if students engaged in activities in which they iterated collections of paces symbolized on paper—the footstrip—instead of measured with units that were parts of their bodies, taking the result of measuring as a given might become taken-as-shared.

Revision 2: The Construction of the Measurement Strip

During one activity in the teaching experiment, two students recorded each iteration of 10 on a piece of adding machine tape that signified a piece of wood that had to be cut 30 cans long (see Figure 6.12). The teacher used this incident to suggest that a piece of paper would be much easier to carry around than the smurf bar, which was 10 connected Unifix cubes, and asked the students to make a strip 50 cans long as a new measurement tool. The students made a strip of 50, but they marked only the decades, as described in Nancy and Meagan's case studies. The reason these students may not have felt the need to make marks for individual cubes may have been that they were making a record of the activity of iterating the entire smurf bar. So when they were measuring with the strip of 50, the students could have intended to use individual cubes to measure the ones, just as they had done with the smurf bar.

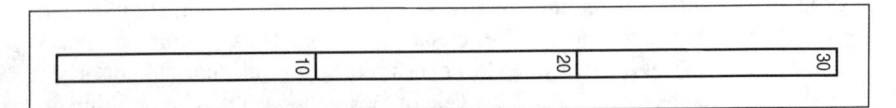

Figure 6.12. Student's record of iterating tens.

The teacher later asked the students to make a strip 10 cans long that could replace the smurf bar, in contrast with the student-initiated tool that arose from the students' efforts to stop iterating the smurf bar. At this point, some of the students did draw marks for the individual cubes because, when they were measuring with the smurf bar, they could measure with either the whole bar or the individual cubes on an as-needed basis. Hence they reasoned that if they were going to replace the bar, they would need marks for individual cubes. This reasoning led naturally to the construction of the measurement strip of 100, which was introduced by taping 10 ten strips together in the context of a discussion on measuring with ten strips.

Revision 3: The Diminished Role for the Measurement Stick

As part of our hypothesized learning trajectory outlined in chapter 4, we anticipated a transition in measuring activity from reasoning with a measuring strip to reasoning with a measuring stick (see Figure 6.13). An analysis of the teaching experiment, however, led us to diminish the role of the measurement stick in the revised teaching experiment. We learned that the measurement stick was not needed to help support the transition from reasoning with the strip to reasoning with the empty number line. One reason was that the measurement stick was not needed to introduce tasks concerning incrementing, decrementing, and comparing. As the students engaged in such activities, they reasoned with the measurement strip in a natural way. In fact, when the measurement stick was introduced, several students had difficulty reading the unit measures. For these reasons, we suggest that its use be limited or eliminated altogether as teachers and researchers initiate a new Design Research Cycle.

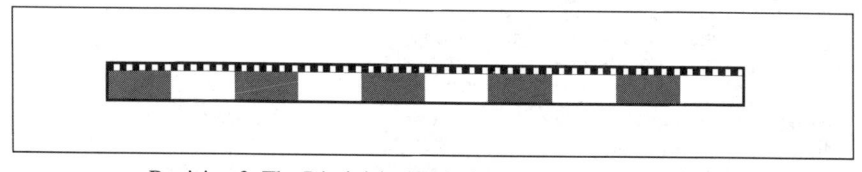

Revision 3: The Diminished Role for the Measurement Stick

Conclusions

Taken together, these three revisions serve as a paradigm case of the feedback mechanism found in the Design Research Cycle. In other words, the daily and retrospective analyses of students' learning in situ provided feedback to inform the designer of revisions to the instructional sequence. After reflecting on these revisions, we were able to create table 6.1, which serves as the outline of our measurement-instruction theory. This table was created using the analyses of classroom mathematical practices and chains of signification as a basis. Specifically, the cascade of tool-use and taken-as-shared purposes of the community as they acted with these tools formed the basis for the first three columns, whereas the analysis of mathematical practice shaped all columns, but mainly the last.

Table 6.1
Overview of the Proposed Role of Tools in the Instructional Sequence

Tool	Imagery	Activity/Taken-as-shared interests	Potential mathematical discourse topics
Feet (heel to toe)		Measuring	
Masking tape	Record of activity of pacing	Reasoning about activity of pacing	Focus on covering distance
Footstrip	Record of pacing (builds on masking tape) (form-function shift: using a record of pacing as a tool for measuring)	Measuring with a "big step" of five = measuring by iterating a collection of paces	Measuring as divorced from activity of measuring; structuring distance in collections of fives and ones
Smurf cans	Stack of Unifix cubes signifies result of iterating	Measuring by creating a stack of Unifix cubes	Builds on measuring divorced from activity of iterating
Smurf bar	Signifies result of iterating	Measuring by iterating a collection of 10 Unifix cubes; structuring distance into measures of tens and ones	Accumulation of distances; coordinating measuring by tens with measuring by ones
10-strip	Signifies measuring tens and ones with the smurf bar	Measuring by iterating the 10-strip, and using the strip as a ruler for the ones	Accumulation of distances
Measurement strip	Signifies measuring with 10-strip; starts to signify result of measuring (form-function shift: inscription developed for measuring is used for scaffolding and communicating)	(1) measuring: strip alongside item; counting by tens and ones ⇒ Reading of endpoint (2) Reasoning about spatial extensions (results of measuring have become entities in and of themselves)	Distance seen as already partitioned; extension already has a measure Part-whole reasoning; quantifying the gaps between two or more lengths Shift in focus: focus on number relations; developing and using emergent framework of number relations
Empty number line	Signifies reasoning with measurement strip	Means of scaffolding and means of communicating about reasoning about number relations	Numbers as mathematical entities (numbers derive their meaning form a framework of number relations) Various arithmetical strategies

GENERALIZATION

At the end of a Design Research Cycle, what can be taken from the experience? In other words, what, if anything, from the measurement classroom teaching experiment generalizes beyond this one enactment of an instructional sequence? In our view, at least two aspects of this study generalize beyond Ms. Smith's first-grade classroom: (1) an instructional theory on linear measurement, and (2) theoretical innovations regarding Design Researching and analyzing learning in social context.

Measurement Instructional Theory

One of the most, if not the most, important products of a series of design experiments is the generation of an instructional theory for a specific mathematical content area that can be used by classroom teachers. After all, one of the main commitments of Design Researchers is to test and revise an instructional theory with the goal of producing a viable theory that can be adapted by classroom teachers. We suggest that the Measurement Instructional Theory outlined in this monograph and summarized in Table 6.1 can be generalized. We are very deliberate when we claim that the instructional *theory,* versus an instructional *sequence,* generalizes. One of the main problems with claiming that a sequence generalizes is that researchers expect that if teachers follow their instructional recipes, they will obtain the same learning results in their classrooms. We believe that no instructional sequence or curriculum materials can be written so that if the teacher follows directions precisely, the teacher will obtain the same results attained in the measurement experiment. The reason for this discrepancy is that classrooms are, by nature, idiosyncratic entities with a wide range of students with diverse backgrounds and abilities. Therefore, we would never expect that the measurement sequence would generalize and produce exactly the same learning results in a variety of classrooms. Rather, we propose that what can be generalized is the measurement instructional *theory*—the ideas embodied in table 6.1—that express the potential tools, discourse, mathematical activity, and imagery that can be integrated to support students' understanding of linear measurement. Individual teachers, who are familiar with the diversity of their particular classrooms, should use the results of our measurement experiment as a guide to develop their own hypothetical learning trajectories specific to the needs of their students. More specifically, we see three aspects that teachers should consider when designing hypothetical learning trajectories for their own classrooms:

1. How the students are expected to act and reason with the tools
2. How this activity relates to the preceding activities
3. The relevant mathematical concepts and the students' conceptual development in relation to them

Our overall goal is that as different teachers develop their own ways of adapting the Measurement Instructional Theory to their classrooms, the results—their own personal instructional theories on measurement—can be shared with others so that they contribute to the overall strengthening of the instructional theory developed in this monograph. In other words, the creating, testing, and revising of measurement instructional theories by other teachers and researchers constitutes another phase of the Design Research Cycle and contributes to the research that began in this measurement teaching experiment.

Theoretical Innovations

A second aspect of this measurement experiment that we contend can be generalized is certain theoretical and practical ideas about designing instruction and analyzing learning in situ. Ideas, such as the methodological approach to analyzing students' learning in context, involving using Toulmin's model of argumentation (chapter 3) and Walkerdine's chains of signification can be used by other researchers who are interested in conducting similar types of analyses. Furthermore, the processes involved in designing instruction, such as developing hypothetical learning trajectories, can be further developed by other researchers who wish to conduct design experiments. Already, several of the authors of this monograph have used the ideas developed in this analysis to design and conduct experiments in differential equations and algebra (Stephan & Rasmussen, in press; Presmeg et al., in press). These new studies show that not only can the ideas in this monograph be used to conduct new research but the new research has served to modify the original theoretical ideas here presented. For example, as a result of our experiences analyzing the mathematical practices from the measurement experiment, the methodology for documenting classroom mathematical practices has now been expanded (see Stephan & Rasmussen, in press). Therefore, the theoretical and practical innovations described in this monograph are already being generalized and used by ourselves and other researchers who are engaged in similar research.

CONCLUSION

In this chapter, we have revisited the three themes of the monograph in terms of the Design Research methodology. First, we made the point that classroom-based analyses, such as those found in chapter 5, that document learning in social context provide teachers with the details for implementing an instructional sequence in their own classrooms. Teachers may draw on both the classroom mathematical practices and the diversity of students' participation in these practices to design hypothetical learning trajectories that fit their classroom situations. Furthermore, we argued that being able to conjecture about 25 individual learning trajectories is not feasible for a teacher and that the classroom mathematical practice provides the teacher a way to support the learning of a community. These types of analyses satisfy two of the three criteria that Cobb outlined in chapter 1, that an analytic approach should make possible the documentation of both the collective mathematical learning and the mathematical reasoning of individual students as they participate in the practices of the community.

Second, we used the notion of a chain of signification to analyze the tool use that coincided with the evolving interests and mathematical practices of the first-grade classroom. We stressed that the chain provides an account of the history of the evolving meaning associated with the formation of signifying links. The chain and associated practices were pragmatically useful in that they provided a backdrop against which to make revisions to the instructional sequence and to elaborate the Measurement Instructional Theory (Table 6.1). We have therefore satisfied Cobb's (chapter 1 of this monograph) criterion that an analytic approach should feed back to inform researchers in the improvement of instructional designs. We presented the revisions and resulting instructional sequence on measuring and arithmetical reasoning in table form, which included a description of each tool along with the imagery, activity, and potential mathematical topics that could possibly arise as the sequences are enacted.

Finally, we argued that the Design Research Cycle, which consisted of successive rounds of design, classroom-based research, and socially situated analyses and revisions, differs from traditional measurement studies. The difference lies in the fact that prior analyses typically focused on pretest and posttest results and seldom tied the processes of learning to the training techniques that were employed. Another way to state this point is that the learning that occurred in the measurement training sessions did not account for the socially situated nature of the experimentation. Thus teachers have few resources or guides for making adjustments to the instruction so that it fits their particular classroom community. One of the strengths of Design Research is that the decisions, whether good or bad, of the research team and the rationale for daily revisions were articulated during the analysis. This reflective process of using students' learning in social context to inform practitioners in their daily instructional decisions and retrospective revisions is the hallmark of this approach.

REFERENCES

Cobb, P. (1999). Individual and collective mathematical learning: The case of statistical data analysis. *Mathematical Thinking and Learning, 1,* 5–44.

Cobb, P., Gravemeijer, K., Yackel, E., McClain, K., & Whitenack, J. (1997). Symbolizing and mathematizing: The emergence of chains of signification in one first-grade classroom. In D. Kirshner & J. A. Whitson (Eds.), *Situated cognition theory: Social, semiotic, and neurological perspectives* (pp. 151–233). Hillsdale, NJ: Erlbaum.

Cobb, P., Stephan, M., McClain, K., & Gravemeijer, K. (2001). Participating in mathematical practices. *Journal of the Learning Sciences, 10*(1, 2), 113–163.

Gravemeijer, K. (1994). *Developing realistic mathematics education.* Utrecht, Netherlands: CD-β Press.

Gravemeijer, K. (1998). Developmental Research as a Research Method. In J. Kilpatrick & A. Sierpinska (Eds.), *Mathematics education as a research domain: A search for identity* (An ICMI study) (book 2, pp. 277–295). Dordrecht, Netherlands: Kluwer.

Gravemeijer, K. (1999). How emergent models may foster the constitution of formal mathematics. *Mathematical Thinking and Learning, 1*(2), 155–177.

Latour, B. (1990). Drawing things together. In M. Lynch & S. Woolgar (Eds.), *Representations in scientific practice.* Cambridge, MA: The MIT Press.

Meira, L. (1998). Making sense of instructional devices: The emergence of transparency in mathematical activity. *Journal for Research in Mathematics Education, 29*(2), 121–142.

Pirie, S., & Kieren, T. (1994). Growth in mathematical understanding: How can we characterize it and how can we represent it? *Educational Studies in Mathematics, 26*, 61–86.

Presmeg, N., Dörfler, W., Elbers, E., van Amerom, B., Yackel, E., Underwood, D., et al. (in press). Realistic mathematics education research: Leen Streefland's work continues. *Educational Studies in Mathematics*.

Saxe, G. (1991). *Culture and cognitive development: Studies in mathematical understanding.* Hillsdale, NJ: Erlbaum.

Stephan, M., & Rasmussen, C. (in press). Classroom mathematical practices in differential equations. *Journal of Mathematical Behavior*.

Walkerdine, V. (1988). *The mastery of reason: Cognitive development and the production of rationality.* London: Routledge.

Chapter 7

Conclusion

Michelle Stephan, *Purdue University Calumet*

The goals of this monograph were both theoretical and pragmatic in nature. Theoretically, we examined three main research questions: (1) How do we characterize learning, in particular, learning to measure, in terms of both social and individual processes? (2) What role do tools play in supporting students' measuring development? (3) How do we account for our presence and influence as instructional designers in the learning process? Each of these research questions was recast as a theme and explored throughout the monograph's chapters. Stephan, in chapter 2, used prior research on students' development of measuring conceptions both to reveal the motivation for further investigation of each theme and to develop our theoretical position on each. Throughout this monograph, the emergent perspective was described as a view that coordinates both social and individual aspects of learning and takes learning with tools as central to mathematical development. The emergent perspective was cast as a view that could expand prior measurement studies that did not take social context or tool use into account during analyses. Further, the heuristics of Design Research were outlined to describe the instructional design theory underlying the research documented in the monograph. Design Research was posed as an alternative to design theories used during prior measurement studies. In the third chapter, we described the methodology that guided the analysis presented in the fifth chapter. The analysis consisted of documenting both the collective and individual learning of the classroom community and the evolving tool use that emerged during the course of the measurement sequence.

Pragmatically, we hoped to extend prior measurement investigations by outlining a potentially viable instructional theory that could be adapted by classroom teachers. In the fourth and sixth chapters, Gravemeijer and his colleagues summarized the process of conjecturing a possible learning trajectory and using analyses of students' learning to revise the measurement sequence both in action and retrospectively. This process of experimentation yielded a potentially viable instructional theory on linear measurement and arithmetic that consisted of the tools that might be used along with the conceptual development that evolves simultaneously. Gravemeijer and others went on to describe how teachers might adapt the sequence to their own students.

In the sections that follow, I conclude the monograph by revisiting and clarifying some important aspects of each of the three themes. I discuss each theme in turn and close the monograph by pointing to future researchable issues.

THEME 1: LEARNING AS A SOCIAL AND INDIVIDUAL PROCESS

The first main theme discussed in the literature review concerned documenting students' learning, in particular, of measurement conceptions, as it develops in social context. From the literature review, I argued that current research advocates building on cognitive analyses of students' measurement development to include a social perspective on learning (cf. Bauersfeld, Krummheuer, & Voigt, 1988; Krummheuer, 2000; Vygotsky, 1978). Rather than swing from one side of the social-versus-individual dichotomy to the other, we advocated a coordination of these two perspectives. In this regard, we were indebted to the pioneers of coordinating social and cognitive perspectives (e.g., Bauersfeld et al., 1988; Cobb and Yackel, 1996). Cobb, Stephan, McClain, & Gravemeijer (2001) describe this coordination best by taking the relation between social and individual perspectives as reflexive. The relationship between these two perspectives is extremely strong in that one does not exist without the other.

The analysis presented in the fifth chapter of this monograph was one example of such a coordination of perspectives. On the one hand, we took a social perspective to document the mathematical interpretations that became taken-as-shared in the public discourse. The teacher's and students' interpretations were seen as constituting that which became normative in the public domain—a social perspective. On the other hand, the case studies presented two students' development—an individual perspective—yet their learning was cast as acts of participation in, and contribution to, the evolving normative activity. Cobb et al. (2001) offer more detail concerning aspects of coordinating these two perspectives. The results, however, of the complementary analysis served as documentation of the evolving classroom mathematical practices and a diversity of ways of participating in them. Coordinating these two perspectives gave a fuller picture of learning than either one of the perspectives could do alone. For example, if we presented only the analysis of the classroom mathematical practices, one might wonder how individual students participated in these practices. Although a mathematical practice is clearly a social construct, we emphasized that students' reasonings were considered as acts of participation in the practice and constituted the reasoning from which the practices arose. Therefore, an analysis of mathematical practices is complemented by individual students' reasoning. If we had documented only individual students' reasoning, the analysis would not have considered the social environment in which these students' learning occurred. As the analysis showed, the reorganizations that these two students made did not occur apart from the classroom conversations in which they participated. In the next sections, I address some remaining concerns about coordinating the two perspectives.

Mathematical Practices

I wish to clarify two main points concerning analysis of mathematical practices. The first concerns the methodological criteria that one adopts when documenting the emergence of mathematical practices. As we detailed in the third and fifth chapters of the monograph, the absence of a backing in students' argumentations was one indication that a practice had become established. This statement does not imply that *any* absence of backings suggests that a practice is taken-as-shared. Mathematical justification, or backings, must be explored publicly *first* before they can drop out of the discourse and indicate a shift in practices. In other words, the absence of discussion about an interpretation does not necessarily imply taken-as-shared understanding. When backings that have been debated, however, publicly drop out of students' argumentations, then we can agree that this absence has significance. A second indication that a practice is relatively stable is that when a student offers a contribution that appears to violate a practice, the other members of the classroom treat the proposal as a breach. Methodologically speaking, this statement implies that the social and sociomathematical norms must also be relatively stable for the researcher to form such conclusions as those mentioned. If students are not obliged to explain their thinking or to challenge one another when a difference in interpretation occurs, then the researcher cannot assume that the absence of backings indicates a shift in taken-as-shared learning.

A second point I wish to clarify concerns the use of a commonly adopted criterion for judging the establishment of a mathematical practice. Under this criterion, a question such as "When do we decide that a mathematical practice is established, when 90 percent of the students share the same view?" is asked. We would argue that such a question imposes an individual perspective on a social phenomenon. To answer the question, one must assess each student's current cognitive understanding and determine whether 90 percent of the students share a sufficiently similar understanding. Our perspective is that a mathematical practice does not assume that everyone "shares" an interpretation. In fact, to claim the establishment of practices makes no claims about how any individual student is reasoning. Rather, the establishment of a mathematical practice means that the community has accepted as taken-as-shared an interpretation for the purposes of continued communication and progression of its members' reasoning. Therefore, we would not consider whether a certain percentage of students shared the same interpretation (see also Bowers and Nickerson, 2001). Instead, we analyze the mathematical reasoning that becomes normative in the public discourse, using the two criteria discussed in the preceding paragraph to determine the collective interpretation that is taken-as-shared.

Case Studies

The two case studies presented in the fifth chapter document different ways of participating in mathematical practices. By documenting these different ways, we hoped to account for the diversity of students' reasoning as they participated in, and contributed to, the local classroom community. A traditional cognitive case study would have documented the shifts in Meagan's and Nancy's reasoning, as well, but this evolution would have been cast purely in terms of their cognitive reorganizations. In other words, traditional case studies would have done well at documenting the shifts that occurred in students' minds but would not necessarily have accounted for the students' development in terms of social interaction. We argue that cognitive analyses can be a useful but incomplete account of students' development. We presented strong evidence in the fifth chapter to suggest that not only did Meagan and Nancy make substantial progress in their thinking, but they did so *as they participated* in their classroom community. Often during the analysis of the mathematical practices, we saw that Nancy and Meagan made important reorganizations during the course of a whole-class discussion or small-group interaction. Therefore, an analysis that documents only students' cognitive reorganizations stops short of explaining the context within which these reorganizations occurred. In this spirit, the case studies were presented alongside, and sometimes within, the analysis of classroom mathematical practices. The practices of the community and individual students' reasoning co-evolved, and presenting the two analyses together was an attempt to illustrate the simultaneity of social and individual learning processes.

The Pragmatics of the Coordination

We have argued theoretically that coordinating social and individual perspectives of learning is essential for understanding students' conceptual development. Pragmatically, such coordination has important consequences. The analysis of the mathematical practices in the fifth chapter not only presented an account of the learning of the classroom community but also outlined the emergence of a potentially powerful instructional theory on linear measurement. The learning detailed in the measuring practice analysis summarized the mathematical content that emerged during the enactment of the sequence of instruction. This analysis was helpful for Gravemeijer and his colleagues, who, in chapter 6, revised the hypothetical learning trajectory to present an instructional theory on linear measurement. Gravemeijer and others were able to draw on the analysis to anticipate both major shifts in collective learning and mathematical topics of conversation that can be initiated by the teacher during the enactment of the measurement sequence.

The case studies are helpful for teachers who want to enact a similar measurement sequence, because the studies detail the contributions made by students as the sequence is enacted. As I have argued before, diversity in students' thinking is crucial for the growth of the community because teachers need to draw from a wide range of contributions for productive discussions. Thus the case studies provide teachers with differing accounts of students' development and help teachers better anticipate students' reasoning that will contribute to their pedagogical agenda.

THEME 2: TOOL USE AS CENTRAL TO MATHEMATIZATION

The literature review in chapter 2 first documented, then challenged prior research on students' development of measuring conceptions and the role of tools in supporting learning. The measurement experiment that is the subject of this monograph was an excellent domain to explore this issue because measuring necessarily incorporates the use of physical devices. Had we started the measurement instructional sequence by presenting students with a conventional ruler and taught them how to use it, the students would have experienced the introduction of this tool in a top-down manner. The tool would have been something separate from their current mathematical experiences, and students would have had to create meaning as they decoded the device. Rather, the approach we took used informal tools to build on students' existing mathematical understanding and led them to progressively construct meaning with several devices prior to the "ruler." In this way, the students developed foundational imagery that grew out of their experience in a bottom-up manner. At the heart of this approach was the conviction that tool use and mathematical meaning develop simultaneously; tool use and meaning making were viewed as reflexively related.

Even though the designer intends for the tools to be experienced in a bottom-up manner, this situation does not always occur. For example, recall from chapter 5 that the first introduction of Unifix cubes as measuring devices was problematic; the task called for reasoning with the new tool in a manner inconsistent with the existing taken-as-shared reasoning. Thus, the analysis of mathematical practices in the fifth chapter and the ensuing chain of signification in chapter 6 were significant in that they described the learning that occurred as students reasoned with tools. We have found an analysis that details the emerging chain of signification to be more helpful than approaches that neglected the role of cultural tools, both for determining the progression of tool use and for documenting the changing interests of the classroom. The chain of signification elaborated in chapter 6 detailed the evolving tool use more explicitly than previous studies were able to do and documented the importance of including several tools that were not originally intended for the instructional sequence.

The chain of signification is important for a second reason. In their chapter, Gravemeijer and his colleagues described the emergent-models heuristic that undergirds the design of an instructional sequence. The backbone of a sequence relies on the idea that students learn as they model their mathematical activity. A sequence is typically organized around a global "model of–model for" shift. As Gravemeijer and others noted, however, the model can take on different manifestations during the realization of an instructional sequence. In the example of the measurement sequence, the "ruler," the overarching model, took several forms, such as the masking-tape record, the footstrip, the smurf bar, the measurement strip, and so on. The chain of signification detailed not only the manifestations and progression of the model during the experiment but also the evolving taken-as-shared meaning and purposes of the classroom community. Therefore, a chain of signification can be a helpful design heuristic, in that the skeleton of a hypothetical learning trajectory can be organized as an anticipated chain of signification. As Table 6.1 indicated, a sequence can be organized according to the anticipated tools that will be used and the imagery and the mathematical activities, interests, and practices that correspond to them.

The analysis of the classroom mathematical practices in the fifth chapter showed that students' learning did not occur apart from reasoning with tools. The chain of signification provided a detailed account of that learning from the perspective of the role that various tools played in students' development of measuring conceptions. This account added to prior measurement studies that did not analyze the role that reasoning with tools plays in supporting learning.

THEME 3: DESIGN RESEARCH

We have argued throughout the monograph that mathematical practices and tools are inseparable aspects of learning. We have also argued that an instructional designer capitalizes both on mathematical practices to form a conjectured, or realized, learning trajectory and chains of signification to anticipate how a model might take on life during instruction. The instructional theory on measurement as detailed by Gravemeijer and his colleagues in the sixth chapter is the product of such analyses and experimentation. The chapters in this monograph are the next iteration in the long cycle of experimentation guided by the Design Research Cycle described in the second chapter by Stephan. The measurement experiment encompasses both phases of the cycle. In the Design Phase, the measurement sequence was prompted by and designed to enhance prior instructional sequences on arithmetical reasoning. A hypothetical learning trajectory was conjectured and subsequently tested in a classroom teaching experiment. The Research Phase was the measurement classroom teaching experiment. The experiment consisted of testing and revising the instructional sequence with a class of students and analyzing the learning that occurred as students reasoned with tools. This analysis then led Gravemeijer and his coresearchers to make revisions to the sequence, the results of which they presented in the sixth chapter, thus completing another iteration of the Design Research Cycle.

That Design Researchers are more than participant observers during experimentation should be clear by now. In fact, we view our role as proactive in shaping the evolving learning that occurs in classrooms. As such, the analyses in this monograph tied students' development to the decisions the researchers made during experimentation. In other words, our accounts of students' learning were tied to the means by which we supported that learning. This type of account provided a clearer picture as to the reasons instruction was changed during and subsequent to experimentation. Thus Design Researchers, one of whom is the classroom teacher, have an important role in shaping the mathematical practices of the classroom community, and this role, from a design perspective, was highlighted and accounted for during the analysis. This type of analysis stands in contrast with prior measurement studies that focused on whether students scored better on tests after measurement training was conducted (see Stephan, chapter 2 of this monograph).

Finally, if we take seriously our contention that learning is inherently social in nature, then we must emphasize that the measurement sequence is not replicable in its purest sense. Social perspectives on learning warn us that each classroom is a different social setting and that an instructional sequence can be realized in a variety of ways. Rather than view this warning as a negative, Gravemeijer and his colleagues argued in chapter 6 not to expect the instructional sequence to be replicable. Instead, they detailed a potential instructional theory that teachers should adapt to their particular setting by formulating their own hypothetical learning trajectories. To aid in this endeavor, Gravemeijer and others detail the mathematical ideas that can arise for students as they engage in the sequence rather than provide teachers with a list of activities or tools that they can use during instruction. The importance of such a description is that it organized instruction on the basis of students' thinking rather than mathematical content and it viewed the teacher as a professional decision maker rather than an invariant of the social context.

UNEXPLORED TERRITORY

Throughout this monograph we have examined a variety of research subjects, including investigating the role that social situations, tools, and designers play in supporting students' mathematical development. In particular, we explored these issues in the context of an investigation in a first-grade classroom that engaged in a measurement instructional sequence. Consequently, we were able to not only examine theoretical ideas but also develop an instructional theory in the domain of linear measurement. For all our ambitious efforts, we are aware that we have underemphasized important aspects of classroom life. For example, although we detailed the decision-making process of the researchers—who included the teacher—during experimentation, the role of the teacher was not examined to its full potential. Whereas powerful mathematical tasks and tools are important for supporting learning, a teacher who understands the mathematical intent of these components is more important. As seen during the analyses in this monograph, the teacher was very skilled at initiating and guiding classroom discourse to bring mathematically sophisticated ideas to the forefront. An analysis of this teacher's pedagogical practices is vital to understanding the way in which the teacher makes on-the-spot decisions that can promote sophisticated discourse and, therefore, learning.

Another area that should be further explored following this study is the role that students' participation in outside cultural practices played in supporting their measuring development. For example, students participate in measuring practices at home or in other outside–of-school environments. Furthermore, we defined *social* at the outset to mean social interactions at the level of the classroom, not as one's social status in the world. An interesting exploration would investigate how students with diverse societal backgrounds participated differently in the classroom mathematical practices. Although these lines of investigations were beyond the scope of this study, we believe they are extremely important aspects of classroom life and that understanding them would foster a stronger measurement instructional theory.

REFERENCES

Bauersfeld, H., Krummheuer, G., & Voigt, J. (1988). Interactional theory of learning and teaching mathematics and related microethnographical studies. In H-G. Steiner & A. Vermandel (Eds.), *Foundations and methodology of the discipline of mathematics education* (pp. 174–188). Antwerp: Proceedings of the TME Conference.

Bowers, J. S., & Nickerson, S. (2001). Identifying cyclic patterns of interaction to study individual and collective learning. *Mathematical Thinking and Learning, 3*, 1–28.

Cobb, P., Stephan, M., McClain, K., & Gravemeijer, K. (2001). Participating in mathematical practices. *Journal of the Learning Sciences, 10*(1&2), 113–163.

Cobb, P., & Yackel, E. (1996). Constructivist, emergent, and sociocultural perspectives in the context of developmental research. *Educational Psychologist, 31*, 175–190.

Krummheuer, G. (2000). Mathematics learning in narrative classroom cultures: Studies of argumentation in primary mathematics education. *For the Learning of Mathematics, 20*(1), 22–32.

Vygotsky, L. (1978). *Mind in society.* Cambridge, MA: Harvard University Press.